VITAL SIGNS

2007–2008

Other Norton/Worldwatch Books

VITAL SIGNS

2007-2008

The Trends That Are Shaping Our Future

WORLDWATCH INSTITUTE

Erik Assadourian, *Project Director*

Molly D. Anderson	Mary Galinski	Lisa Mastny
Elroy Bos	Gary Gardner	Esmeralda Meyer
Michael Browne	Brian Halweil	Ishani Mukherjee
Katie Carrus	Alana Herro	Danielle Nierenberg
Zoë Chafe	Lindsay Hower Jordan	Shyama Pagad
Sean Charles	Suzanne Hunt	Rodrigo G. Pinto
Alessandra Delgado	Kai N. Lee	Michael Renner
Kevin Eckerle	Nicholas Lenssen	Janet Sawin
Christopher Flavin	Ling Li	Peter Stair
Hilary French	Yingling Liu	Elanor Starmer

Linda Starke, *Editor*
Lyle Rosbotham, *Designer*

W.W. Norton & Company
New York   London

people causing this destruction have names and addresses. They include you and me and all the other consumers in the world. They include politicians who make empty promises (or no promises at all). They include corporate executives who continue to ignore the realities of doing business on a finite and fragile planet and instead put profit over long-term concerns (not to mention spending billions to encourage consumers to crave things that are bad for them and the planet).

Each of these individuals has opportunities—some big, some small—to become less of a destructive influence and more of a restorative force. They can trade their car for a bicycle or bus pass, they can invest responsibly, they can eat less meat, start an urban farm, use less energy, green their businesses, lobby their governments, and teach others about these problems and how to implement solutions in their own communities. There are thousands of changes we can make—many painless, some rewarding, and others challenging. But to do nothing will lead to tragedy.

Now let me thank some of the people who helped with our humble attempt to bring wider attention to the trends shaping our future. This year's edition of *Vital Signs* had many helpers: in writing the book, assembling it, communicating our findings, and of course funding it.

Let me start with our funders—a group of foundations, governments, and international agencies who along with our members and subscribers keep the Institute running. Thanks especially to the Blue Moon Fund, Ecos Ag-Basel, the Energy Future Coalition and the Better World Fund, the Ford Foundation, the government of Germany, The Goldman Environmental Prize, the Richard & Rhoda Goldman Fund, the W. K. Kellogg Foundation, the Steven C. Leuthold Family Foundation, the Noble Venture Fund of the Community Foundation Serving Boulder County, the Norwegian Royal Ministry of Foreign Affairs, the Natural Resources Defense Council, the Prentice Foundation, the V. Kann Rasmussen Foundation, the Rockefeller Brothers Fund, The Shared Earth Foundation, The Shenandoah Foundation, the Taupo Community Fund of the Tides Foundation, the United Nations Environment Programme, the Food and Agriculture Organization of the United Nations, the United Nations Population Fund, the Wallace Genetic Foundation, Inc., the Wallace Global Fund, the Johanette Wallerstein Institute, and the Winslow Foundation. We also wholeheartedly thank the Worldwatch Board of Directors and the more than 3,500 Friends of Worldwatch for their dedication to the Institute and the issues we address. Their generosity of time and resources makes our work possible each year.

Of course, a huge thank you goes to all the Worldwatch staff who contributed an article or more to this year's book. This *Vital Signs* also benefited from a broad array of expertise beyond the Institute. Former Worldwatch researcher Nicholas Lenssen once again tracked nuclear energy trends. Former Worldwatch MAP fellow Peter Stair described the declining state of male reproductive health. Katie Carrus, former Worldwatch intern now at the US Humane Society, provided an analysis of global egg production. Kevin Eckerle, Congressional Fellow of the American Association for the Advancement of Science, investigated climate change's effect on biodiversity. Lindsay Hower Jordan, of Rare (an international conservation organization), summarized current HIV/AIDS trends. Alessandra Delgado of the Public Health Institute contributed an overview of the informal economy. Molly D. Anderson of Food Systems Integrity and Elanor Starmer of the Friedman School of Nutrition Science and Policy at Tufts University addressed the issue of growing concentration in the agricultural sector. Mary Galinski and Esmeralda Meyer of the International Center for Malaria Research and Education at Emory University provided an overview of malaria transmission. And finally, we also included two articles on biodiversity from the World Conservation Union (IUCN). Thanks to Elroy Bos for his article on endangered species and to Shyama Pagad and Michael Browne for their article on invasive species.

Worldwatch's current accomplished group of interns also helped a tremendous amount.

First, a special thanks to Sean Charles, *Vital Signs* intern who along with analyzing soybean production reviewed the articles for data and citation accuracy—a vital role. Worldwatch intern Ishani Mukherjee coauthored the article on fossil fuel trends and intern Rodrigo G. Pinto coauthored the piece on biofuels. Thanks also to interns Una Song, who helped to compile data for the energy articles, and Marine Confavreux, who assisted with the one on solar power.

To guarantee that Worldwatch provides the best analysis and most up-to-date information, we depend on many reviewers and experts to provide us with data. Thanks to all who helped ensure a strong edition of *Vital Signs* this year, including: Julian-G. Albert, Linda Allen, Christoph Berg, Raffaella Bianchin, Travis Bradford, Nancy Chege, Diana Leafe Christian, Brec Cooke, Colin Couchman, Neeraj Doshi, Pat Franklin, Uwe R. Fritsche, Peter Gleick, Michael Greger, Peter Raymond Johnson, Egil Juliussen, Christian Kjaer, Anna Knee, Jennifer Lacquet, Craig Long, Birger Madsen, Eric Martinot, Andrew McMullin, Timothy Newman, Miyun Park, Steven Piper, Sandra Postel, Paul Racionzer, Payal Sampat, Pengfei Shi, Niels Skakkebæk, Vladimir Slivyak, Markus Steuer, Shanna Swan, John Talberth, Dick Urban Vestbro, Jean-Christophe Vié, Mathis Wackernagel, Carolin Wahnbaeck, Carol Welch, Philip White, Angelika Wirtz, and Paul Zajac.

We are grateful, too, for the continuing efforts of W. W. Norton & Company, and for the help provided by Amy Cherry, Leo Wiegman, and Anna Oler. It is their commitment that transforms *Vital Signs* from computer files to the volume you're reading and gets it to bookstores and classrooms across the United States.

We are also lucky enough to have a committed group of international partners who work hard to produce *Vital Signs* outside the United States. For their extensive assistance in translating, publishing, and promoting recent editions, we thank Soki Oda of Worldwatch Japan, Anastàsia Monjas at Centro UNESCO de Catalunya in Spain, and Eduardo Athayde in Brazil.

Vital Signs depends as well on the efforts of many individuals at Worldwatch working long hours behind the scenes. Patricia Shyne, our Publications Director, works with our publisher and international partners to make this a truly international book. Our development staff—Georgia Sullivan, Mary Redfern, Laura Parr, and Courtney Berner—play a critical role in cultivating support for Worldwatch's essential work. Behind the scenes, we receive daily support from Director of Finance Barbara Fallin and from Mail and Publication Fulfillment Coordinator Joseph Gravely—who, sadly, is retiring this year after 18 years of tireless service.

Our communications team—Darcey Rakestraw and Julia Tier—work diligently to bring Worldwatch publications to new audiences every day. Molly O'Meara Sheehan—when not busy traveling the world to launch *State of the World 2007: Our Urban Future*, which she was Project Director of—found time to help with the article on urbanization. Senior Editor Lisa Mastny helped to make sure that all the t's were crossed and i's dotted this year, which we are grateful for. And *World Watch* Magazine Editorial Director Tom Prugh kept us inspired with powerful new stories every few months.

At the heart of producing *Vital Signs* are two very important individuals. Linda Starke, an independent editor who has worked on Worldwatch publications for 25 years, is the linchpin who holds the *Vital Signs* project together. Worldwatch's Art Director Lyle Rosbotham brings the book its artistry—from the beautiful cover shot of Bear Glacier to the crisp color scheme and clean layout throughout.

All in all, this effort took hundreds of individuals and thousands of supporters to make it happen. Thanks to every one of them. But remember, *Vital Signs* can only inform. It is up to each of us to let this information change us and inspire us to in turn make change.

Erik Assadourian
Project Director

Worldwatch Institute
1776 Massachusetts Ave., N.W.
Washington, DC 20036

TECHNICAL NOTE

Units of measure throughout this book are metric unless common usage dictates otherwise. Historical population data used in per capita calculations are from the Center for International Research at the U.S. Bureau of the Census. Historical data series in *Vital Signs* are updated in each edition, incorporating any revisions by originating organizations.

Unless otherwise noted, references to regions or groupings of countries follow definitions of the Statistics Division of the U.N. Department of Economic and Social Affairs.

Data expressed in U.S. dollars have for the most part been deflated to 2006 terms. In some cases, the original data source provided the numbers in deflated terms or supplied an appropriate deflator. Where this did not happen, the U.S. implicit gross national product deflator from the U.S. Department of Commerce was used to represent price trends in real terms.

Preface

Our latest assessment of the world's "vital signs" reveals an important new element of the "sustainability" crisis now unfolding. Global energy and food markets have collided over the past year, greatly increasing pressure on the renewable resources that nourish the economy. This collision between two of the world's largest and most essential economic sectors will have complex repercussions. One consequence is clear: unprecedented stress on Earth's land and water resources will present difficult choices for policymakers for a long time to come.

Rising incomes and changing diets in China and other Asian countries have increased the demand for livestock products. Global meat production rose 2.5 percent to 276 million tons in 2006, which in turn has increased the consumption of corn, soybeans, and other agricultural commodities used as animal feed. China's imports of soybeans from Brazil doubled between 2004 and 2006, driven by a combination of the growing need for animal feed and falling water tables in northern China, where most of its soybeans are grown.

At the same time, three years of high oil prices, combined with growing concern about energy security and climate change, have fostered a booming market for biofuels—up 28 percent in 2006 alone. The United States is now the world's largest consumer of ethanol fuel, most of it produced from corn, the country's most abundant crop. In 2006, 16 percent of the U.S. corn crop went into ethanol production

and corn prices nearly doubled as a result.

High oil prices, advancing technologies, and strong political support are expected to increase demand for biofuels for many years. The world's automobile and truck fleet now numbers nearly 900 million, while the biofuels produced in 2006 were sufficient to run no more than 10 million. The United States and Brazil will lead the way in expanding that production in the years ahead, but scores of other countries are planning to introduce incentives for the use of biofuels.

The ecological risks of rising food and energy demand became more apparent in 2006. Palm oil—which is used for cooking and, more recently, as a supplement to diesel fuel—became a hot commodity in 2006, spurring entrepreneurs to clear tropical forests in Southeast Asia in order to expand their palm plantations.

In the United States, the search for more land to grow corn—the total corn acreage is projected to rise 15 percent in 2007—may cut into the Conservation Reserve, a federal program designed to protect erosion-prone soils from cultivation. And as U.S. farmers switched from planting soybeans to planting corn, the price of soybeans also rose, encouraging further expansion of Brazil's rapidly growing soybean farms. Brazil is one of the few countries whose agricultural frontier continues to spread, and the biologically rich grasslands and forests on the southern edge of the Amazon are now being cleared to grow soybeans and other crops.

The energy and food economies are colliding on many different fronts, but fossil fuel–driven climate change may be the most profound. The Intergovernmental Panel on Climate Change reported in early 2007 that global warming may undermine agricultural productivity in many regions—just when the need to replace fossil fuels increases demand on agricultural resources. And if a combination of climate change and forest clearing eventually destroys the Amazon forest, the rainfall that nourishes some of the world's most productive food and energy crops in central Brazil could be greatly diminished.

Vital Signs 2007–2008 highlights some of the early responses that could help bring the food and energy economies into more sustainable balance. Changes in agricultural practices and consumption patterns, for instance, are urgently needed, since today's agriculture is highly inefficient in its use of energy and resources. No-till cropping and reduced meat consumption, to cite two examples, could go a long way toward improving the sustainability of agriculture.

On the energy front, rising prices have begun to spur investment in energy efficiency and in a host of renewable energy technologies, including wind power, which was up 26 percent in 2006, and solar power, up 40 percent. And more-sustainable approaches to biofuels production are also under development; a growing number of companies are investing in technologies that can produce biofuels from agricultural wastes and perennial grasses that not only have a lower environmental impact but can actually increase the amount of carbon stored in soils. Changes in government incentives will be needed, however, if the biofuels industry is to make this transition before serious damage is done to the world's forests and agricultural lands.

The converging food and energy markets are also beginning to have economic and social impacts. From the feedlots of Kansas to the tortilla markets of Mexico, people are complaining about the rising price of corn. Food inflation is expected to accelerate. And while the impact may hardly be noticeable in industrial countries, poor consumers in developing countries will not be so fortunate.

We point out in the pages that follow that the world is making progress—but still has a long way to go—in reducing poverty and achieving the other U.N. Millennium Development Goals. Although the proportion of people suffering from hunger worldwide has declined modestly in the last decade, the number of chronically malnourished people has risen to more than 800 million.

It may seem surprising, but rising agricultural prices can have positive impacts as well. Under the North American Free Trade Agreement, Mexican farmers have suffered from a flood of cheap, subsidized U.S. corn. Higher prices may save some of Mexico's small farms and villages, which have been rapidly losing population due to a weak agricultural economy.

The energy industry is well known for its tendency to generate great wealth for those who own fossil fuels while leaving most people behind. Whether energy derived from agricultural resources will provide wider economic benefits to society at large will depend on the distribution of land and the structure of agriculture. These in turn depend on public policy decisions that must be made soon. As one of this year's "vital signs" points out, large-scale traders and processors increasingly dominate agricultural markets and reap a disproportionate share of the profits. That trend needs to be reversed if the world's poor are to benefit from rising prices for agricultural products.

The collision of the food and energy economies is another reminder of the powerful forces that connect the human economy to Earth's ecological systems. Efforts to replace limited fossil fuels are placing new demands on biological resources—even as the continued combustion of those fuels further weakens the resource base by disrupting the climate. In the dangerous period we've now entered, Planet Earth's vital signs require careful monitoring.

Christopher Flavin
President
Worldwatch Institute

VITAL SIGNS

2007–2008

Part One

KEY INDICATORS

Food and Agriculture Trends

Center-pivot irrigation in Colorado, United States

Tim McCabe/Natural Resources Conservation Service

▶ Grain Production Falls and Prices Surge

▶ Soybean Demand Continues to Drive Production

▶ Meat Output and Consumption Grow

▶ Seafood Increasingly Popular and Scarce

▶ Irrigated Area Stays Stable

For data and analysis on food and agricultural trends, including sugar consumption, pesticide trade, and fertilizer use, go to www.worldwatch.org/vsonline.

Grain Production Falls and Prices Surge

Brian Halweil

In 2006, world grain production dropped to 1,994 million tons—a fall of about 55 million tons, or some 2.7 percent, from the previous year.[1] (See Figure 1.)

Economists, hunger activists, and agricultural researchers track world grain production because people still primarily eat foods made from grain. On average, humans get about 48 percent of their calories from grains, a share that has declined just slightly from 50 percent over the last four decades.[2] Grains, particularly maize (corn) in conjunction with soybeans, also form the primary feedstock for industrial livestock production.

Global grain production per person dropped from 318 kilograms in 2005 to 305 kilograms in 2006.[3] (See Figure 2.) But output per person varies dramatically by region. For instance, it stands at roughly 13,000 calories per day in the United States, most of which is fed to livestock, compared with 2,700 calories in China and just 670 calories in Zimbabwe.[4] (One kilogram of grain contains about 3,500 calories.)

Production of the three major grain crops—wheat, corn, and rice—all declined in 2006, as the world's major growing areas suffered poor weather.[5] Wheat output in 2006 stood at roughly 592 million tons, down almost 33 million tons—5.3 percent—from 2005.[6] This is the largest reduction since 1994 and was provoked by severe drought and heat across Australia and in Europe's wheat belt, as well as unseasonably cold dry weather during planting time in North America and the Black Sea region.[7] Global stocks of wheat declined by 16 percent since 2005, corn stocks were down nearly 20 percent, and total stocks, including rice, dropped by 17 percent.[8] (See Figure 3.)

Typhoons, drought, flooding, diseases, and insect attacks marred the 2006 rice crop across Asia.[9] Global production fell to 421 million tons, slightly down from 422 million tons in 2005.[10] In India, the 2006 monsoon season, which ended in September, was erratic; several important rice-producing states, such as Assam, Tamil Nadu, and Uttar Pradesh, received less than the normal amount of precipitation, while rainfall was above average in Orissa.[11] Crops in China were also affected by droughts, floods, and disease problems, which kept production nearly the same as in 2005 despite larger plantings.[12]

The world corn crop in 2006 was estimated at 694 million tons, 2.2 percent below the previous year, due to smaller crops in Argentina, South Africa, and the United States—which alone is responsible for 40 percent of the global crop.[13]

At the same time, global demand for corn jumped due to the rapid expansion in corn-based ethanol production, primarily in the United States. The amount of corn used for ethanol there has grown from just 6 percent of domestic production in 2000 to an estimated 20 percent in 2006, or roughly 55 million tons, about the same amount as is exported.[14] There are currently 110 ethanol plants operating in 20 states across the country, with 79 additional plants under construction, which will more than double national capacity.[15]

By late 2006, rising demand combined with the poor grain harvests in key producing nations to push world grain prices to their highest levels in a decade.[16] In November, the U.S. hard wheat export price averaged $219 a ton, up about one third from the previous year.[17] The U.S. export price for No. 2 yellow maize averaged $164 per ton, up about 70 percent from the previous year.[18]

In the case of corn and wheat, these high prices will likely encourage farmers to plant more land in crops in 2007. But most analysts suspect that if the use of corn for ethanol continues to grow at the current pace, it may take more than one good crop season for prices to retreat significantly from their current highs.[19]

According to the latest *Food Outlook* from the U.N. Food and Agriculture Organization, global expenditures on imported foodstuffs in 2006 could reach a historic high of $383 billion, more than 2 percent above the previous year's level.[20] Import bills for developing countries are expected to have been almost 5 percent higher in 2006, mainly as a result of price increases rather than an increase in the actual volume of food imports.[21] Higher prices will force many countries to cut back on food imports.

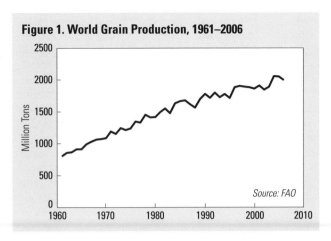

Figure 1. World Grain Production, 1961–2006

Source: FAO

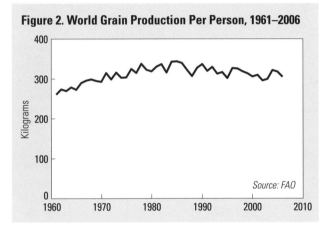

Figure 2. World Grain Production Per Person, 1961–2006

Source: FAO

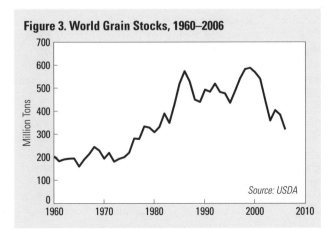

Figure 3. World Grain Stocks, 1960–2006

Source: USDA

World Grain Production, 1961–2006

Year	Total	Per Person
	(million tons)	(kilograms)
1961	805	261
1965	914	273
1966	992	290
1967	1032	296
1968	1065	299
1969	1073	295
1970	1087	293
1971	1194	315
1972	1156	299
1973	1246	316
1974	1216	303
1975	1241	304
1976	1348	325
1977	1333	315
1978	1454	338
1979	1413	323
1980	1418	319
1981	1496	331
1982	1552	337
1983	1478	316
1984	1632	343
1985	1665	344
1986	1678	340
1987	1618	323
1988	1565	307
1989	1700	328
1990	1779	337
1991	1717	320
1992	1798	330
1993	1727	313
1994	1777	317
1995	1715	302
1996	1883	327
1997	1903	326
1998	1891	319
1999	1882	314
2000	1861	306
2001	1909	310
2002	1844	296
2003	1891	300
2004	2055	322
2005	2049	318
2006 (prel)	1994	305

Source: FAOSTAT.

Soybean Demand Continues to Drive Production   *Sean Charles*

The world soybean harvest reached a record 214 million tons in 2005, the latest year with data, an increase of 4.4 percent from 2004.[1] (See Figure 1.) The United States, Brazil, Argentina, and China accounted for 90 percent of that output.[2] (See Figure 2.)

The United States is the largest producer of soybeans, with an output of 83.4 million tons in 2005.[3] Over the past 25 years, however, its market dominance has eroded.[4] The United States produced 60 percent of the world's soybeans in 1980 but only 39 percent in 2005.[5] The country's declining role as an exporter can be traced to increased competition with South American producers, growing domestic competition with corn, the production of biodiesel, and the resistance in some markets to genetically modified (GM) soybeans.[6]

Soybeans enrich the soil with nitrogen, which can then be used by other plants, making them beneficial for crop rotations.[7] In the United States, this has usually meant planting soybeans and corn in alternating years. But high demand for corn for ethanol production and distiller's grains (a high-protein animal feed) has driven many farmers to plant two years of corn for every year of soybeans.[8] This in large part explains the 7-percent decline in total U.S. soybean harvested area in 2005.[9] Globally, however, harvested area stayed stable at 92 million hectares.[10] (See Figure 3.)

Brazil produces a quarter of the soybeans worldwide and in 2003 became the largest exporter.[11] Its success in this field is largely due to vast tracts of undeveloped land.[12] The 11 states of the center-west and Amazonia regions, which include the *cerrado*—the world's most diverse savanna—and large portions of the Amazon rainforest, doubled production from 2000 to 2005.[13] Production in Argentina is growing even faster, with an increased output of 216 percent since 1995.[14] Rapid South American soybean expansion is creating mono-crop plantations at a rate that endangers 22 million hectares of tropical forest and savanna in the next 20 years.[15]

Global growth in wealth and in industrial agriculture has resulted in greater consumption of meat and convenience foods, raising demand for soybeans as animal feed and as soybean oil (the most widely used vegetable oil).[16] Soybean meal, the protein-rich solid produced in the soybean crushing and oil extraction process, accounts for 65 percent of the world's protein feed.[17] The majority of soy meal is used for animal feed, including 98 percent in the United States.[18]

Increased reliance on soy meal for industrial agriculture to supply China's huge and increasingly urban population, coupled with the growing scarcity of agricultural land, has made China reliant on imported soybeans.[19] Even though soybean cultivation began in China 5,000 years ago, in 2005 the country imported 74 percent of its soy.[20] After entering the World Trade Organization in 2002, China reduced trade restrictions and doubled its imports to 21.4 million tons in 2003—accounting for 55 percent of its consumption.[21] Soy meal demand in China and in Southeast Asia is reliant on poultry production, so success in controlling avian flu is expected to lead to further demand increases.[22]

Genetically modified soybeans were introduced to the market in 1996 to be resistant to the pesticide glyphosate, commonly sold as Roundup.[23] In 2005, "Roundup Ready" soybeans accounted for 87 percent of the crop in the United States and 98 percent in Argentina. Similarly, GM soybeans accounted for 41 percent of Brazil's harvested area—an 88 percent increase from 2004.[24] Though the European Union was the top soy meal importer in 2005, it imports very little soybean oil for human consumption because of mandatory GM labeling and public stigma surrounding genetic engineering.[25]

Sustained demand increases are expected for soybeans for animal feed, vegetable oil, and biodiesel, with a projected growth of 60 percent by 2025.[26] In 2005, soybean oil accounted for 92 percent of the 250 million liters of biodiesel made in the United States, a recent use that is bound to grow as Americans turn to biofuels to replace imported oil.[27] Similarly, 59 percent of Brazilian biodiesel came from soy.[28]

LINKS pp. 24, 40

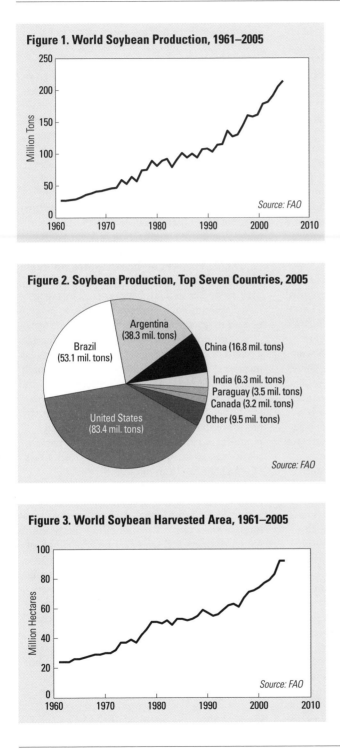

Figure 1. World Soybean Production, 1961–2005

Source: FAO

Figure 2. Soybean Production, Top Seven Countries, 2005

Argentina (38.3 mil. tons)

Brazil (53.1 mil. tons)

China (16.8 mil. tons)

India (6.3 mil. tons)
Paraguay (3.5 mil. tons)
Canada (3.2 mil. tons)
Other (9.5 mil. tons)

United States (83.4 mil. tons)

Source: FAO

Figure 3. World Soybean Harvested Area, 1961–2005

Source: FAO

World Soybean Production, 1961–2005

Year	Production
	million tons
1961	27
1965	32
1966	36
1967	38
1968	41
1969	42
1970	44
1971	46
1972	47
1973	59
1974	53
1975	64
1976	57
1977	74
1978	75
1979	89
1980	81
1981	89
1982	92
1983	79
1984	91
1985	101
1986	94
1987	100
1988	94
1989	107
1990	108
1991	103
1992	114
1993	115
1994	136
1995	127
1996	130
1997	144
1998	160
1999	158
2000	161
2001	178
2002	181
2003	191
2004	205
2005	214

Source: FAO.

Meat Output and Consumption Grow

Danielle Nierenberg

In 2006, meat production increased 2.5 percent to an estimated 276 million tons; output is expected to rise another 3 percent in 2007 to 285 million tons.[1] (See Figure 1.) Developing countries produced more meat and dairy products than industrial countries for the first time in 1995.[2] At least 60 percent of meat in 2006 was produced in developing nations.[3]

Consumption of meat and other animal products also continues to grow. While 43 kilograms of meat are produced annually per person at the moment, however, meat consumption still varies widely by region and socioeconomic status.[4] In the developing world, people eat about 32 kilograms of meat per year, compared with almost 85 kilograms per person in the industrial world.[5] (See Figure 2.)

Due mainly to the spread of avian flu and the culling of birds and burning of meat, global poultry output rose only slightly in 2006 to approximately 83 million tons, roughly a 1-percent increase from the preceding year.[6] Beef output rose by 2.5 percent, with nearly 66 million tons produced in 2006.[7] The United States is the largest beef producer, although developing nations account for 55 percent of the total.[8]

Pork production grew by 3 percent to 108 million tons, more than any other meat.[9] (See Figure 3.) This increase is likely due to shifting meat consumption patterns in Asia as people switch from chicken to pork due to concerns about avian flu.[10] China continues to be the world's largest producer of pig meat, but several South American nations, including Brazil—which accounts for nearly 70 percent of pork output in the region—as well as Chile and Mexico, are increasing their production facilities.[11]

Much of the growing demand for animal products is being met by large-scale intensive industrial systems (factory farms).[12] These facilities rely on commercial breeds of livestock, usually pigs and chickens, that have been bred to gain weight quickly on soybeans and corn. Factory farms are very crowded and confine animals in close quarters. Many of the world's 17 billion hens and meat chickens are given an area less than the size of a sheet of paper to live

LINKS
pp. 22, 42, 86, 90

in, while cattle in feedlots often stand knee-deep in manure and arrive at slaughterhouses covered in feces.[13]

These operations are increasingly located in or near urban markets in developing countries, making cities the center of industrial meat production. Although city dwellers have kept livestock for centuries to help deal with urban waste as well as to provide income and food, industrial operations can create a host of environmental and public health problems, including the spread of diseases such as avian flu.[14]

Livestock are also the "single largest anthropogenic user of land," according to the Food and Agriculture Organization.[15] Meat production accounts for 70 percent of all agricultural land and 30 percent of the land surface of the planet.[16] In the Amazon, 70 percent of previously forested land is occupied by pastures for cattle and much of the remaining 30 percent is used to grow soybeans and other feed crops.[17]

In addition, livestock are responsible for 18 percent of greenhouse gas (GHG) emissions (as measured in carbon dioxide equivalent), which is higher than the share contributed by cars and sport utility vehicles.[18] And livestock account for 37 percent of emissions of methane, which has more than 20 times the global warming potential of carbon dioxide, and for 65 percent of nitrous oxide, another powerful GHG, most of which comes from manure.[19]

Livestock are major water users and polluters as well. The irrigation of feed crops for cattle accounts for nearly 8 percent of global human water use.[20] Compounding the contamination of rivers and streams from the runoff of manure from feedlots, livestock waste can contaminate soil and groundwater with a cocktail of hormones, pesticides, and antibiotics used in factory farms.[21] One way to prevent some of these problems is by raising livestock in areas with enough land to handle the waste from large operations. Thailand, for example, has levied high taxes on poultry production within a 100-kilometer radius of Bangkok while exempting farmers outside that zone.[22] Over the last decade, poultry production near Bangkok dropped significantly.[23]

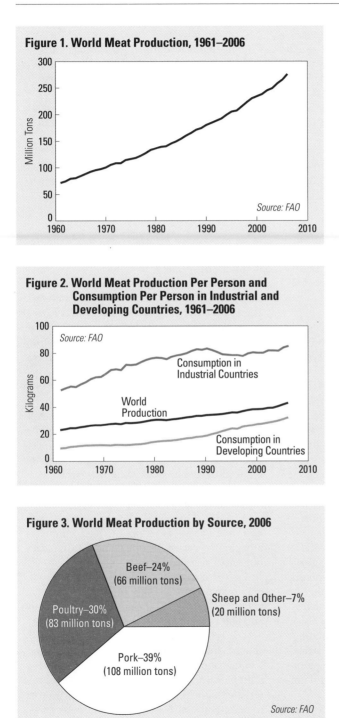

Figure 1. World Meat Production, 1961–2006

Source: FAO

Figure 2. World Meat Production Per Person and Consumption Per Person in Industrial and Developing Countries, 1961–2006

Source: FAO

Consumption in Industrial Countries

World Production

Consumption in Developing Countries

Figure 3. World Meat Production by Source, 2006

Beef–24% (66 million tons)

Sheep and Other–7% (20 million tons)

Poultry–30% (83 million tons)

Pork–39% (108 million tons)

Source: FAO

World Meat Production, 1961–2006

Year	Total	Per Person
	(million tons)	(kilograms)
1961	71	23.1
1965	84	25.2
1966	88	25.7
1967	92	26.4
1968	95	26.7
1969	97	26.7
1970	100	27.1
1971	105	27.6
1972	108	27.9
1973	108	27.5
1974	114	28.4
1975	116	28.3
1976	118	28.5
1977	122	28.9
1978	127	29.6
1979	133	30.3
1980	136	30.7
1981	139	30.7
1982	140	30.4
1983	145	30.9
1984	149	31.3
1985	154	31.8
1986	160	32.4
1987	165	32.8
1988	171	33.5
1989	174	33.5
1990	180	34.1
1991	184	34.3
1992	188	34.5
1993	192	34.8
1994	199	35.4
1995	205	36.0
1996	207	35.9
1997	215	36.9
1998	223	37.7
1999	230	38.3
2000	234	38.4
2001	238	38.6
2002	245	39.3
2003	249	39.5
2004	258	40.6
2005	269	41.9
2006 (prel)	276	43.0

Source: FAO.

Seafood Increasingly Popular and Scarce

Brian Halweil

People around the world ate about 156 million tons of seafood in 2004, the last year for which there are data.[1] This is a relatively large jump from the preceding year—almost 9 million tons—about half of which was satisfied by a rebound in certain wild fish populations, with the other half representing continued rapid growth in fish farming.[2] (See Figure 1.) (Since seafood is generally consumed fresh or within a few months of being caught, statistics on consumption and production are nearly identical.)

Since 1950, seafood consumption has jumped almost eight times.[3] This rise in global consumption comes even as seafood becomes scarcer. In 2006, scientists tracking historical changes in the world's major fish populations estimated that all major fish stocks could be commercially extinct—less than 90 percent of their historic levels—by the middle of this century if current trends continue.[4]

On average, each person ate three times as much seafood in 2004 as in 1950 (see Figure 2)—but the amount and type of seafood consumed vary widely.[5]

LINKS p. 100

The Chinese consume about a fifth of the world's seafood, eating per person roughly five times as much seafood as they did in 1961.[6] Total Chinese fish consumption has increased more than 10-fold in that time.[7] (See Figure 3.) Over the same period, U.S. seafood consumption jumped 2.5 times.[8] The Japanese consume the most seafood per person, about 66 kilograms each year.[9] In Europe, the average person eats about 26 kilograms a year, slightly more than the average Chinese does.[10]

For people in wealthy nations, seafood is an increasingly popular health food option; given its high levels of fatty acids and trace minerals, nutritionists recognize it as essential to the development and maintenance of good neurological function, not to mention reduced risk of cancer, heart disease, and other debilitating conditions.[11] In poorer nations in Asia, Africa, and Latin America, people are also eating more fish, if they can afford it or can fish for it themselves.[12] For more than 1 billion people, mostly in Asia, fish supply 30 percent of the protein they consume, compared with just 6 percent worldwide.[13]

Consumers in Europe, the United States, and Japan favor larger, predatory fish, like tuna and cod, the populations of which are most endangered.[14] Most salmon and shrimp, two other popular items, are now raised in farms that use several times more fish as feed than they actually produce.[15]

In contrast, poorer people tend to depend on smaller fish that are lower on the food chain, including herbivorous farmed fish like catfish, carp, and tilapia, as well as oysters, clams, mussels, and sea vegetables.[16] In China, which raises 70 percent of the world's farmed fish, fish farming accounts for nearly two thirds of total fish consumption and is dominated by such herbivorous species.[17]

With the depletion of wild fish schools, virtually all of the growth in the global catch today comes from farmed fish.[18] Whereas wild harvests have stagnated over the last 10 years, fish farming's output has more than doubled to 59.4 million tons, accounting for nearly 40 percent of the global harvest.[19]

Although farmers have been raising herbivorous fish in ponds for millennia, the relatively recent move toward raising tuna, salmon, striped bass, shrimp, and other carnivores in pens consumes a growing share of the world's fish. Species like anchovy, herring, capelin, and whiting are reduced to feed for animals or fish farms. In 1948, only 7.7 percent of total landings turned into fishmeal and fish oil.[20] Currently, 37 percent of global marine landings—about 32 million tons a year—is reduced to feed, eliminating an important historical and future source of human sustenance.[21]

Fish also sustain people as a livelihood, employing about 38 million people worldwide.[22] Of these, 95 percent are smaller fishers and fish farmers in Asia and Africa.[23] Smaller vessels employ more people per ton of fish caught, and they also wield more exacting and less damaging fishing tools—hand lines rather than nets dragged across the bottom—a characteristic that will be important as countries try to maintain their fishing communities even as there are fewer fish to catch.[24]

Figure 1. World Fish Harvest, 1950–2004

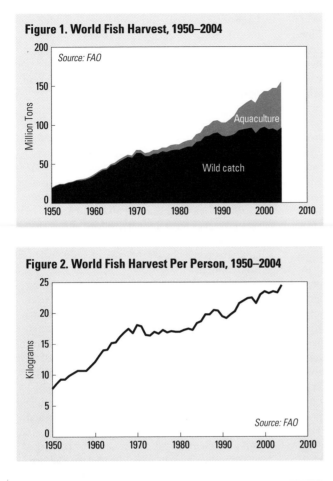

Source: FAO

Aquaculture

Wild catch

Figure 2. World Fish Harvest Per Person, 1950–2004

Source: FAO

Figure 3. Seafood Consumption in Top Four Countries or Regions, 1961 and 2003

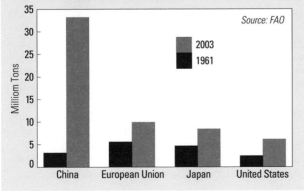

Source: FAO

2003
1961

China   European Union   Japan   United States

World Fish Catch and Aquaculture, 1950–2004

Year	Catch	Aquaculture
	(million tons)	
1950	19	1
1955	27	1
1960	35	2
1965	49	3
1970	64	4
1971	64	4
1972	60	4
1973	60	5
1974	63	5
1975	63	5
1976	66	5
1977	65	6
1978	67	7
1979	68	7
1980	68	7
1981	70	8
1982	72	8
1983	72	9
1984	78	10
1985	79	11
1986	85	13
1987	86	14
1988	89	16
1989	90	16
1990	86	17
1991	85	18
1992	86	21
1993	88	24
1994	93	28
1995	94	31
1996	95	34
1997	96	36
1998	89	39
1999	95	43
2000	97	46
2001	94	49
2002	95	52
2003	92	55
2004	96	59

Source: FAO.

Irrigated Area Stays Stable

Ling Li

In 2003, the latest year with data, the world's irrigated area stagnated at about 277 million hectares—150,000 hectares less than in 2002.[1] (See Figure 1.) The annual expansion rate of world irrigated land has fallen from an average of over 2 percent from 1961 to 1992 to around 1 percent from 1993 to 2003.[2] Irrigated area per person in 2003 stood at 0.044 hectares, 7 percent below the 1978 peak.[3] (See Figure 2.)

Irrigation is distributed unevenly around the world: 42 percent of arable land in Asia was irrigated in 2000, while in sub-Saharan Africa the figure was only 4 percent.[4] Irrigated areas in China, India, Pakistan, and the United States account for half of the world total, but China and India each have almost 20 percent.[5] (See Figure 3.) In 1992, India passed China as the country with the most irrigated land.[6]

Irrigated area accounted for only 20 percent of total arable land in 1997–99, yet farmers worldwide harvested 40 percent of all crops and nearly 60 percent of cereals from these lands.[7] In China, the 45 percent of agricultural land that is irrigated produced 75 percent of the nation's food in 2002.[8] By 2030, 70 percent of world cereal grains will come from irrigated land.[9]

LINKS pp. 20, 22, 42, 86

The worldwide slowdown in irrigation expansion has several sources. For one, it is becoming more expensive for farmers and governments to put in new irrigation because of rising costs of investment in irrigation systems and difficulties developing new sites.[10] But the two major constraints on expansion are soil salinization—a particularly acute problem in semiarid areas when salts build up in the soil as irrigation water evaporates—and shortages of irrigation water, which are driven by both aquifer depletion and competition for water.[11] An estimated 20–30 million hectares of world irrigated land have been degraded by the accumulation of salts.[12]

The overpumping of groundwater for irrigation is now a widespread problem.[13] The number of tubewells supplying underground water to irrigated land has grown rapidly in the last 40 years in India, China, Pakistan, Mexico, and many other countries.[14] Groundwater levels in large areas in India and China are estimated to drop 1–3 meters each year, allowing saltwater to intrude into aquifers, raising pumping costs, and causing land subsidence.[15]

Agriculture accounted for nearly 70 percent of the world's use of fresh water in 2000, although in Asia and the Pacific region the figure was as high as 90 percent.[16] Nevertheless, more irrigation water continues to be transferred to nonfarm uses because of the rapidly growing demands of industries and cities.[17] Yet in China, irrigation's share of total water use dropped from 85 percent in 1980 to 64 percent in 2005.[18] More municipal and industrial wastewater is reused for irrigation, although this raises significant environmental and health concerns when the wastewater receives little or no treatment.[19]

Climate change also threatens irrigation by shifting world rainfall patterns, changing river flows, raising sea levels, and intensifying hurricanes and monsoons.[20] Irrigated areas that rely on water from mountain snowmelt are at particular risk.[21] In South Asia, accelerated glacial melt and reduced rainfall pose problems for the major local crops, such as paddy rice and wheat.[22]

More than half of the irrigation water removed from rivers and aquifers disappears before benefiting a crop, either wasted through evaporation and inefficient irrigation practices or recharged to groundwater.[23] In Asia, the widespread use of pump irrigation is believed to create natural incentives for farmers to be more careful in water management, as they have to pay for energy even though the water is free.[24]

Low-cost treadle pumps and small mechanical pumps have been introduced in South Asia and Africa to help poor farmers get access to irrigation.[25] Drip irrigation is a more efficient technology than flooding and sprinklers, reducing water use by 30–70 percent and increasing crop yields by 20–90 percent.[26] In Kenya, a type of bucket drip irrigation kit costing some $15 has been used for irrigation of small plots of vegetables and fruit trees, generating monthly revenue of about $20.[27] Planting more water-efficient grains can also help reduce water use.[28]

Figure 1. World Irrigated Area, 1961–2003

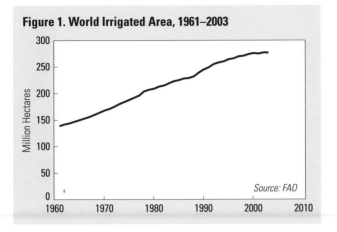

Source: FAO

Figure 2. World Irrigated Area Per Thousand People, 1961–2003

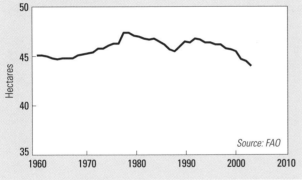

Source: FAO

Figure 3. Irrigated Area in Selected Countries, 1961–2003

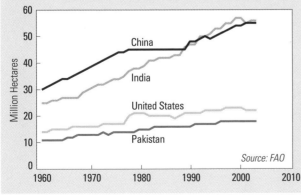

China

India

United States

Pakistan

Source: FAO

World Irrigated Area, 1961–2003

Year	Area
	(million hectares)
1961	139
1962	142
1963	144
1964	147
1965	150
1966	153
1967	156
1968	160
1969	164
1970	168
1971	171
1972	175
1973	180
1974	184
1975	188
1976	192
1977	196
1978	204
1979	207
1980	209
1981	213
1982	215
1983	219
1984	223
1985	225
1986	228
1987	229
1988	232
1989	239
1990	245
1991	249
1992	255
1993	258
1994	260
1995	264
1996	266
1997	270
1998	271
1999	274
2000	276
2001	275
2002	277
2003	277

Source: FAO.

Energy and Environment Trends

Energy delivery in a Mumbai slum, especially important during the monsoon when firewood is soaked.

▶ Fossil Fuel Use Up Again

▶ Nuclear Power Virtually Unchanged

▶ Wind Power Still Soaring

▶ Solar Power Shining Bright

▶ Biofuel Flows Surge

▶ Carbon Emissions Continue Unrelenting Rise

▶ Weather-related Disasters Climb

▶ Ozone Layer Stabilizing But Not Recovered

For data and analysis on energy and climate trends, including hydropower and energy productivity, go to www.worldwatch.org/vsonline.

Fossil Fuel Use Up Again

Janet L. Sawin and Ishani Mukherjee

Global oil use rose by 1 percent in 2006, down from a 1.6-percent increase in 2005 and a near-record 3.9-percent rise in 2004.[1] Global oil consumption reached 3.9 billion tons in 2006.[2] (See Figure 1.)

North America and Asia remain the world's leading oil users, at 25.3 million barrels and 21.4 million barrels a day in 2006, respectively.[3] The United States drained 20.7 million barrels of oil daily—24 percent of the global total.[4] Yet U.S. gasoline use dropped by about 1 percent from the previous year as consumers reacted to higher prices.[5] Other top consumers include Europe at 16.1 million barrels daily, China at 7.2 million barrels a day, and the Middle East at 6.5 million barrels daily.[6]

Oil prices rose for the fourth consecutive year due to continued production declines in many countries and political developments that have slowed output in some nations.[7] Prices averaged $62 per barrel in 2006, up from $58 in 2005.[8] (See Figure 2.) They peaked in July, at $78.40 per barrel, not far from the inflation-adjusted record price of $87 per barrel in 1981.[9]

World oil production also grew by 1 percent in 2006—led by Africa, where new oil fields continue to be developed.[10] Significant new discoveries were made in Brazil, the Middle East, and the Gulf of Mexico.[11] U.S. production leveled off after more than a decade of decline.[12] However, production declines continued for many major producers, including Mexico, Norway, the United Kingdom, Nigeria, Indonesia, and Venezuela.[13] The rapid run-up in Russian production that marked the early years of this decade has now slowed.[14] Iran saw a slight rise in output, but analysts say the country could become a net importer within a decade, due to stagnant production and soaring domestic demand.[15]

LINKS pp. 40, 42, 44

Natural gas and coal data for 2006 are not yet available, but consumption of both fuels rose in 2005.[16] (See Figure 3.) Natural gas use rose by 2 percent that year, to 2.5 billion tons of oil equivalent.[17] North America was the only region to see a decline (down 1.5 percent).[18] Coal use was up by 4.7 percent, to 2.9 billion tons of oil equivalent, with most of the growth in China, which used 1,082 million tons.[19] The next largest consumers were the United States at 575 million tons and India at 213 million tons of coal burned in 2005.[20] The United States accounted for about one fifth of world coal use in 2005.[21]

Some 150 new coal plants, representing almost 100 gigawatts of capacity, could come online in the United States by 2030.[22] Concerns about climate change will likely stop many of these, but at least a dozen are already under construction.[23] India, which accounts for less than 8 percent of global coal use, could also see significant growth.[24] Coal demand there is expected to quadruple by 2031 to sustain economic growth of 9 percent annually, requiring a major rise in coal imports.[25]

China is second only to the United States in total energy use.[26] In 2006, China's energy use rose 9.3 percent—with coal demand up 9.6 percent, crude oil use up 7.1 percent, and natural gas use up 20 percent.[27] China imported 47 percent of its oil in 2006.[28] Increasing demand for oil is driven mainly by the rise in private motor vehicles.[29] It took China nearly 20 years to have 10 million vehicles in 2003 but only three more years to double that number.[30] China's electric generating capacity rose more than 20 percent in 2006—to 622,000 megawatts—and by some estimates a large coal-fired power plant comes online there weekly.[31]

The International Energy Agency projects that, if left unchecked, global energy use will rise more than 50 percent by 2030, with fossil fuels remaining the dominant energy source.[32] In turn, vulnerability to price shocks and supply disruptions would rise, and carbon dioxide (CO_2) emissions could increase by more than 50 percent.[33] Other scenarios project that energy efficiency improvements and renewable energy could displace a significant share of fossil fuel use and reduce global emissions.[34] In March 2007, the European Union committed to reducing CO_2 emissions 20 percent and increasing renewable energy to 20 percent of total energy use by 2020.[35]

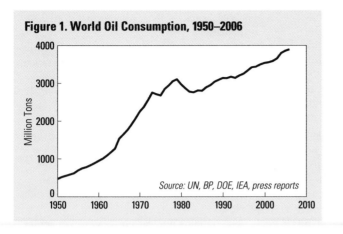

Figure 1. World Oil Consumption, 1950–2006

Source: UN, BP, DOE, IEA, press reports

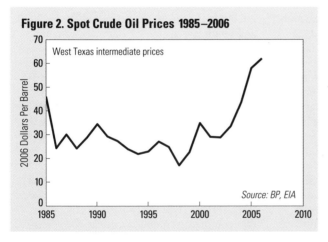

Figure 2. Spot Crude Oil Prices 1985–2006

West Texas intermediate prices

Source: BP, EIA

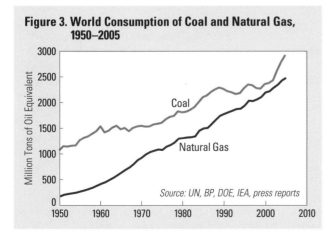

Figure 3. World Consumption of Coal and Natural Gas, 1950–2005

Coal

Natural Gas

Source: UN, BP, DOE, IEA, press reports

World Fossil Fuel Consumption, 1950–2006

Year	Oil	Natural Gas	Coal
	(million tons of oil equivalent)		
1950	470	171	1,074
1955	694	266	1,270
1960	951	416	1,544
1965	1,530	632	1,486
1970	2,254	924	1,553
1975	2,678	1,075	1,613
1976	2,852	1,138	1,681
1977	2,944	1,169	1,726
1978	3,055	1,216	1,744
1979	3,103	1,295	1,834
1980	2,972	1,304	1,814
1981	2,868	1,318	1,826
1982	2,776	1,322	1,863
1983	2,761	1,340	1,916
1984	2,809	1,451	2,011
1985	2,801	1,493	2,107
1986	2,893	1,504	2,143
1987	2,949	1,583	2,211
1988	3,039	1,663	2,261
1989	3,088	1,738	2,293
1990	3,136	1,774	2,270
1991	3,134	1,806	2,225
1992	3,170	1,836	2,203
1993	3,139	1,869	2,168
1994	3,204	1,877	2,186
1995	3,252	1,938	2,282
1996	3,335	2,032	2,353
1997	3,421	2,025	2,338
1998	3,437	2,059	2,283
1999	3,499	2,103	2,273
2000	3,537	2,192	2,361
2001	3,555	2,215	2,381
2002	3,589	2,286	2,434
2003	3,656	2,342	2,629
2004	3,799	2,425	2,799
2005	3,859	2,475	2,930
2006 (prel)	3,896	n.a.	n.a.

Source: UN, BP, DOE, IEA, press reports.

Nuclear Power Virtually Unchanged

Nicholas Lenssen

Between 2005 and 2006, total installed nuclear generation capacity increased by less than 1,000 megawatts, or 0.2 percent, to roughly 370,000 megawatts.[1] (See Figure 1.) Over the past decade, global nuclear power capacity has grown by less than 1 percent on average, far below the pace of renewable energy resources and fossil fuels.[2]

The slight increase in capacity in 2006 was due to the opening of two new reactors, one each in China and India, and to capacity increases at existing reactors in the United States and some European countries.[3] These additions were offset, however, by eight small reactors that were permanently shut—four in the United Kingdom, two in Bulgaria, and one each in the Czech Republic and Spain.[4] Overall, 124 reactors totaling 36,800 megawatts have been closed after an average lifespan of 21 years.[5] (See Figure 2.)

LINKS pp. 42, 80

Still, construction started on six reactors in 2006, three of which are in China.[6] The combined 5,288 megawatts in capacity of new construction projects was at the highest level in a decade.[7] (See Figure 3.)

Globally, only 26 reactors for a combined capacity of 19,778 megawatts are currently under active construction.[8] More than a dozen countries, however, are planning to add new reactors—but as the past has shown, moving from discussions to successful construction can prove difficult. Indeed, over the last 45 years U.S. utilities have ordered at least 120 reactors that they later cancelled—more than the total number operating in that country now.[9]

At the end of 2006, utilities and private consortia had plans to construct some 30 reactors in the United States.[10] But even the leading nuclear producers are uncertain whether these will be built. Duke Energy's chief executive officer testified that he was increasingly pessimistic about his company's chances of building its two planned reactors on schedule—if at all—due to cost and nuclear waste concerns.[11] And the largest nuclear operator, Exelon Corporation, says it will not build a new plant until the waste issue is resolved.[12] The bottom line is that Standard & Poor's Ratings Services does not expect

a new U.S. reactor to be operational until 2015 at the earliest.[13]

In Europe, just one reactor is under construction, in Finland. But in 2006 its builders twice announced delays for its completion, moving it from 2009 to 2011.[14] Areva and the Finnish utility are in a dispute over who will pay cost overruns, which now reach some €1 billion.[15]

A French company plans to start construction in 2007 on the first new reactor in France in eight years.[16] Meanwhile, Prime Minister Tony Blair faces opposition within his own party to his call for the United Kingdom to add reactors.[17]

Asia remains nuclear power's growth center: China plans to add some 32,500 megawatts by 2020; India expects 16,500 megawatts by 2020; and South Korea plans 9,200 megawatts by 2016.[18] Japan only had two new reactors under construction at the end of 2006.[19] Asian countries already accounted for more than 70 percent of the total capacity of new nuclear reactors being constructed at the end of 2006; their share of new nuclear activity is likely to increase in the next five years as China and other countries start new projects.[20]

Even in Asia, however, not all projects are progressing uninterrupted. The international consortia building two reactors in North Korea formally cancelled the project shortly before that country's testing of a nuclear weapon in 2006 and subsequently requested that North Korea pay $1.9 billion in compensation for the aborted project.[21]

The United Nations Security Council called on Iran to cease its uranium enrichment activities and approved the imposition of economic sanctions on the country despite claims that its nuclear effort was for peaceful purposes.[22] Russia, though, continued to help Iran move closer to completing that country's first commercial reactor by the end of 2007.[23] At the same time, in what could be seen as a double standard, the U.S. Congress approved a nuclear cooperation agreement between the United States and India even though India refuses to sign the Nuclear Non-Proliferation Treaty and has nuclear weapons.[24]

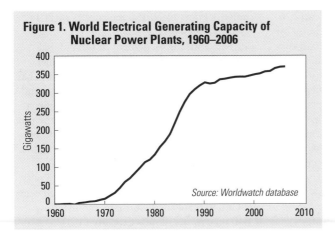

Figure 1. World Electrical Generating Capacity of Nuclear Power Plants, 1960–2006

Source: Worldwatch database

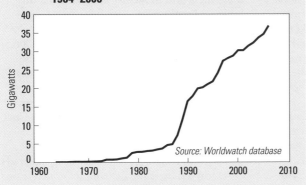

Figure 2. Nuclear Capacity of Decommissioned Plants, 1964–2006

Source: Worldwatch database

Figure 3. World Nuclear Reactor Construction Starts, 1960–2006

Source: Worldwatch database

World Net Installed Electrical Generating Capacity of Nuclear Power Plants, 1960–2006	
Year	Capacity
	(gigawatts)
1960	1
1965	5
1970	16
1971	24
1972	32
1973	45
1974	61
1975	71
1976	85
1977	99
1978	114
1979	121
1980	135
1981	155
1982	170
1983	189
1984	219
1985	250
1986	276
1987	297
1988	310
1989	320
1990	328
1991	325
1992	327
1993	336
1994	338
1995	340
1996	343
1997	343
1998	343
1999	346
2000	349
2001	352
2002	357
2003	358
2004	366
2005	369
2006	370

Source: Worldwatch Institute database, IAEA, and press reports.

Wind Power Still Soaring

Janet L. Sawin

Global wind power capacity increased nearly 26 percent in 2006, exceeding 74,200 megawatts by year's end.[1] (See Figure 1.) The almost 15,200 megawatts of new capacity added in 2006 was 32 percent above the previous year's record.[2] (See Figure 2.) New wind installations were second only to natural gas–fired power plants for the seventh consecutive year in Europe and for the second year in a row in the United States.[3] Wind power remains one of the world's fastest-growing energy sources, driven by concerns about climate change and energy security.[4]

The United States led the world in new installations for the second year running.[5] (See Figure 3.) The addition of 2,454 megawatts pushed total U.S. capacity above 11,600 megawatts—trailing only Germany and Spain.[6] The increase came despite the military's concerns about possible turbine interference with radar signals, which put the installation of thousands of megawatts on hold in several states.[7] An extension of the federal tax credit through 2008 and renewable electricity mandates in more than 20 states drove the surge.[8] Texas breezed past California to become the nation's top wind power generator.[9]

LINKS p. 42

European wind capacity rose 19 percent in 2006, with new records in several countries.[10] Installations of nearly 7,600 megawatts pushed the region's total above 48,000 megawatts, enough to meet more than 3 percent of European Union electricity demand in an average wind year.[11]

Germany saw an upsurge in 2006 with the addition of 2,233 megawatts, an increase of almost 24 percent over 2005.[12] It remains clearly in the global lead, with total wind power capacity exceeding 20,600 megawatts.[13] Renewable energy (mostly wind power) accounted for about 11.6 percent of Germany's electricity market by the end of 2006.[14]

Spain is hanging on to second place (barely ahead of the United States), with total capacity of 11,615 megawatts.[15] An estimated 1,587 megawatts of wind capacity were added in 2006.[16] New installations were 10 percent below the previous year's additions, and there is concern that Spain is not on track to reach the government's target of 20,155 megawatts by 2011.[17]

Although Germany, Spain, and the United States account for 59 percent of total global installations, the wind industry is no longer reliant on just a handful of markets.[18] More than 50 nations tap the wind to produce power, and 13 countries now have more than 1,000 megawatts installed.[19] European countries in the top 10 list in 2006 included France, Denmark, Portugal, the United Kingdom, and Italy.[20]

Outside of Europe, Asia experienced the strongest growth in 2006, adding nearly 3,680 megawatts of wind capacity.[21] India trailed only the United States and Germany by installing 1,840 megawatts of new capacity in 2006.[22] In total installations, India remains in fourth place, with 6,270 megawatts.[23]

China is rapidly catching up, however. It leads the world in the use of small wind turbines and ranks sixth overall for total wind power installations, not far behind Denmark.[24] China added nearly 1,350 megawatts in 2006, thanks to a new renewable energy law, more than doubling its total capacity to 2,604 megawatts.[25] The government plans to redouble its wind capacity by 2010—a target some experts believe will be greatly exceeded—and to install 30,000 megawatts of capacity by 2020.[26]

Globally, wind installation costs have increased in recent years due to rising materials costs (driven by rising demand for concrete and steel) and a general shortage of wind turbines.[27] In the United States, costs have risen over 50 percent since 2003, yet wind remains competitive as costs are rising for all power technologies.[28] Oil giants BP and Shell have joined some of the biggest players in wind power in the United States and elsewhere, accelerating the trend of large corporations investing in wind and other renewable energy technologies.[29]

Investments in new wind power generating equipment exceeded $20 billion in 2006 and are projected to surpass $60 billion in 2016.[30] The Global Wind Energy Council forecasts that with strong policies in place, global installed wind capacity will reach 135,000 megawatts by 2010 and could exceed 1 million megawatts by 2020.[31]

Figure 1. World Wind Energy Generating Capacity, 1980–2006

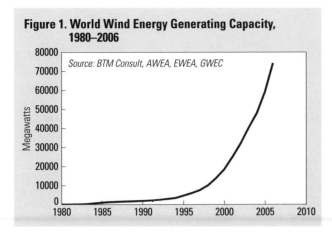

Figure 2. Annual Additions to World Wind Energy Generating Capacity, 1980–2006

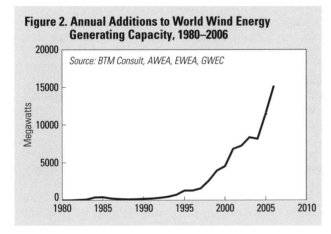

Figure 3. Annual Wind Energy Additions in Germany, the United States, and Spain, 1980–2006

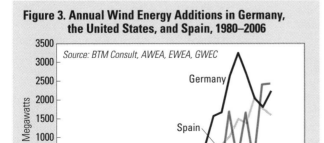

World Wind Energy Generating Capacity, Total and Annual Additions, 1980–2006

Year	Total	Annual Additions
	(megawatts)	
1980	10	5
1981	25	15
1982	90	65
1983	210	120
1984	600	390
1985	1,020	420
1986	1,270	250
1987	1,450	180
1988	1,580	130
1989	1,730	150
1990	1,930	200
1991	2,170	240
1992	2,510	340
1993	2,990	480
1994	3,490	730
1995	4,780	1,290
1996	6,070	1,290
1997	7,640	1,570
1998	10,150	2,600
1999	13,930	3,920
2000	18,450	4,500
2001	24,930	6,820
2002	32,040	7,230
2003	40,300	8,340
2004	47,910	8,150
2005	59,091	11,492
2006 (prel)	74,223	15,197

Source: BTM Consult, AWEA, EWEA, GWEC.

Solar Power Shining Bright

Janet L. Sawin

Global production of photovoltaic (PV) cells—which produce electricity directly from sunlight—rose to a record 2,521 megawatts in 2006, an increase of 41 percent over the previous year.[1] (See Figure 1.) New grid-based installations increased by an estimated 47 percent, to a record 2,000 megawatts, making solar PV the world's fastest-growing energy source.[2] Since 2000, annual global production is up six-fold.[3] (See Figure 2.)

Japan remains the leading PV manufacturer, with a 37-percent market share and production of 927 megawatts of solar cells in 2006.[4] (See Figure 3.) Yet production there grew a lackluster 11 percent in 2006, compared with 38 percent in 2005, and Japan's market share is down from a high of 50 percent two years earlier—with much of this difference yielded to China.[5] Japan's new installations, however, were up 20 percent, to an estimated 350 megawatts of new PV capacity installed.[6]

Germany continues to dominate the market—with roughly 1,100 megawatts added, it accounted for more than half of all new installations in 2006.[7] Germany's additions rose more than 46 percent from 2005, another record-setting year.[8] Europe—led by Germany—continues to rank second in production, with PV manufacturing up 42 percent (to 678 megawatts) in 2006.[9]

LINKS p. 42

The big surprise of 2006 was the dramatic growth in production in China and Taiwan, which now rank third and fifth respectively.[10] In 2003 these two countries manufactured a total of only 26 megawatts. Yet in 2006 they produced an estimated 547 megawatts, accounting for almost half of the global expansion in output and nearly 22 percent of the market.[11] China's leading PV manufacturer, Suntech Power, climbed from the world's eighth largest producer in 2005 to fourth in 2006.[12] Most of China's production was for export to Germany and Spain, with only 25 megawatts installed domestically in 2006.[13]

U.S. production rose 31 percent in 2006, to nearly 202 megawatts, but the country still fell to fourth place behind China.[14] Most of this increase came from one company, First Solar,

which produces an innovative thin film cell that requires less polysilicon than traditional solar cells.[15] Its competitors were constrained by polysilicon supply problems.[16] The United States ranked third behind Germany and Japan for installations, with an estimated 100 megawatts connected to the grid—an increase of 60 percent over 2005.[17] Most of this new capacity was added in California and New Jersey, driven by strong state policies combined with a federal tax credit and rising energy costs.[18]

Spain added about 75 megawatts to come in fourth for new installations.[19] The market there is expected to pick up dramatically thanks to a new building code that requires all new large nonresidential buildings to generate a portion of their electricity with PVs.[20]

The significant increase in global production in 2006 came despite a reported shortage of polysilicon, which slowed growth in some of the world's largest firms.[21] For the first time ever, more than half of the world's polysilicon supply was used for solar cells.[22] Enough new production capacity is now being developed that some analysts predict excess polysilicon capacity within the next few years.[23]

The shortage of polysilicon is also driving advances in thin films, which increased from 6 to 7 percent of the world market in 2006 and could achieve a 20 percent share by 2010.[24] The United States leads the world in thin films, with more than half of world production in 2006.[25] The Japanese automaker Honda plans to open a new thin film plant in 2007.[26]

Exceptionally strong growth in demand for PV and a bottleneck in polysilicon supply have combined to drive up prices over the past two years.[27] Average module prices were $3.85 per watt in late 2006, up from $3.50 per watt in 2005.[28] The U.S.-based Prometheus Institute projects, however, that as production costs fall, technologies continue to advance, and supply and demand come into balance, PV prices will fall more than 40 percent in the next three years relative to prices in late 2006.[29] Such a decline would make solar electricity far more affordable in markets across the globe.

Figure 1. World Annual Photovoltaic Production, 1980–2006

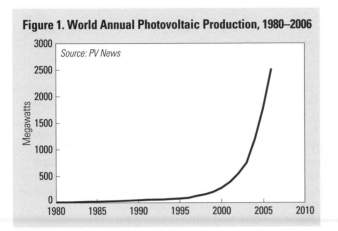

Source: PV News

Figure 2. World Cumulative Photovoltaic Production, 1980–2006

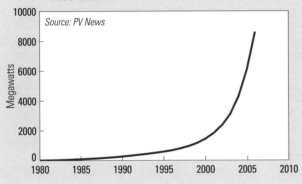

Source: PV News

Figure 3. Photovoltaic Production, Selected Countries and Europe, 1994–2006

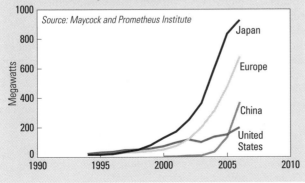

Source: Maycock and Prometheus Institute

Japan
Europe
China
United States

World Photovoltaic Production, 1980–2006

Year	Annual Production	Cumulative Production
	(megawatts)	
1980	7	19
1981	8	27
1982	9	36
1983	17	53
1984	22	75
1985	23	98
1986	26	124
1987	29	153
1988	34	187
1989	40	227
1990	47	273
1991	55	329
1992	58	387
1993	60	447
1994	69	516
1995	78	594
1996	89	682
1997	126	808
1998	155	963
1999	201	1,164
2000	277	1,441
2001	386	1,827
2002	547	2,374
2003	748	3,122
2004	1,194	4,316
2005	1,782	6,098
2006 (prel)	2,521	8,619

Source: PV News.

Biofuel Flows Surge

Rodrigo G. Pinto and Suzanne C. Hunt

World production of biofuels rose 28 percent, to 44 billion liters, in 2006.[1] (See Figure 1.) Production of fuel ethanol—derived mainly from sugar or starch crops—increased 22 percent, passing 38 billion liters.[2] Production of biodiesel, the other major biofuel (made from vegetable oils or animal fats), increased by 80 percent, topping 6 billion liters.[3] In 2006, the surge in production of these two biofuels accounted for 17 percent of the increase in supply of all liquid fuels worldwide, and together they constituted nearly 1 percent of the global liquid fuel supply.[4]

Having passed Brazil in 2005, the United States extended its lead as the largest producer of fuel ethanol.[5] (See Figure 2.) Despite hefty trade barriers, U.S. imports of fuel ethanol increased more than sixfold in order to meet a 35-percent increase in U.S. demand.[6] Brazil in turn boosted its ethanol exports by 71 percent.[7] The United States and Brazil combined produced nearly 90 percent of the world's fuel ethanol.[8] The booming U.S. industry is projected to double its production capacity by the end of 2008.[9]

Germany kept its wide lead in biodiesel production and capacity in 2006, but the United States, France, Italy, and smaller suppliers gained ground.[10] (See Figure 3). The spread of biodiesel production was propelled by especially rapid growth in Malaysia, China, Colombia, Brazil, the Philippines, and the United States.[11]

LINKS pp. 22, 32, 42, 54

The main forces driving this expansion include high oil prices, the use of ethanol in place of toxic fuel additives such as MTBE and lead, mounting concerns about climate change, and a growing array of government mandates and incentives that have strong support from the agriculture sector.[12] Energy security and foreign policy goals provided additional motivation for policymakers in the wake of rising tensions and instabilities involving several major petroleum suppliers.[13]

Responding to these incentives, investment in biofuels soared.[14] In the United States, $14 billion was invested in ethanol stocks in the 12 months through mid-2006, and venture capital-

ists put $813 million into U.S. biofuels during the year as a whole.[15] Much of the venture capital focus has been on developing and commercializing technologies that can break down cellulosic plant material so that producers can move beyond the food crops relied on thus far.[16] In Brazil, $10 billion was invested in the ethanol industry in 2006.[17]

Biofuel assets experienced a roller-coaster year, with booming demand partially offset by volatile grain and oil markets.[18] Growing demand pushed up the price of a wide range of agricultural commodities in 2006, including sugar, corn, soybeans, rapeseed, and palm oil, which in turn cut into the profit margins of biofuel producers.[19]

Concern about the social and environmental impacts of biofuel crops also began to raise questions about the fast-growing industry.[20] Efforts to develop international sustainability standards and assurance systems for biofuels intensified.[21] Germany instituted the first legally binding sustainability system, and others are expected to follow.[22] Increased investment in dedicated sources of biofuels such as jatropha biodiesel and compressed biomethane and in new fuels such as biobutanol and biokerosene started to move the industry in new directions.[23]

Dozens of governments made or reinforced biofuel commitments in 2006, including Argentina, Australia, China, Germany, Italy, Myanmar, the Philippines, South Africa, and South Korea.[24] California, Minnesota, and New York in the United States and New South Wales and Queensland in Australia made stronger biofuel commitments than those of their own national governments.[25]

China, India, South Africa, and the European Commission joined Brazil and the United States to form the International Biofuels Forum, which seeks to expand the biofuels market.[26] In addition, a number of governments and partners created the Global Bioenergy Partnership to promote sustainable bioenergy.[27] And 14 African governments joined Senegal in founding the Pan-African Non-Petroleum Producers Association, aimed in part at building a robust biofuels industry.[28]

Figure 1. World Biofuel Production, 1975–2006

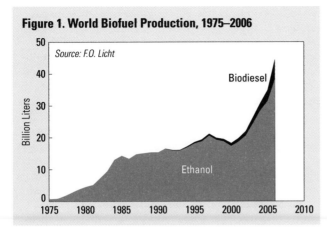

Source: F.O. Licht

Figure 2. Ethanol Production, United States and Brazil, 1975–2006

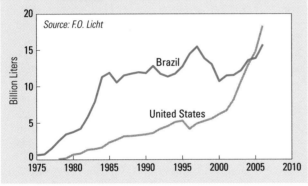

Source: F.O. Licht

Figure 3. Biodiesel Production, Top Four Nations, 2002–06

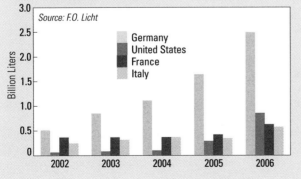

Source: F.O. Licht

Germany
United States
France
Italy

World Ethanol Production, 1975–2006, and Biodiesel Production, 1991–2006

Year	Ethanol	Biodiesel
	(million liters)	
1975	556	
1976	664	
1977	1,470	
1978	2,529	
1979	3,533	
1980	4,368	
1981	4,977	
1982	7,149	
1983	9,280	
1984	12,880	
1985	14,129	
1986	13,193	
1987	14,599	
1988	14,902	
1989	15,191	
1990	15,190	
1991	16,348	11
1992	15,853	88
1993	15,873	143
1994	16,857	283
1995	18,066	408
1996	18,750	546
1997	20,517	570
1998	19,203	587
1999	18,711	719
2000	17,279	872
2001	18,655	1,063
2002	20,529	1,323
2003	24,459	1,846
2004	28,464	2,245
2005	31,340	3,415
2006 (prel)	38,200	6,153

Source: F.O. Licht

Carbon Emissions Continue Unrelenting Rise   Janet L. Sawin

In 2006, atmospheric carbon dioxide (CO_2) concentrations reached 381.84 parts per million by volume, an increase of 0.6 percent over the record set in 2005.[1] (See Figure 1.) Average CO_2 concentrations have risen 20.8 percent since measurements began in 1959 and are now more than 100 parts per million higher than in pre-industrial times.[2]

Fossil fuel burning represents about 80 percent of this increase.[3] In 2005, the last year with relevant data, carbon emissions from this source increased 3 percent to 7.56 billion tons—more than one ton for every person on Earth.[4] Annual emissions from fossil fuels have risen 17 percent just since 2000.[5] (See Figure 2.)

The United States remains the world's top emitter, accounting for over 21 percent of carbon emissions from fossil fuel burning in 2005.[6] U.S. carbon emissions are still on the rise, but growth rates slowed in 2005 to 0.8 percent, down from a 1.7-percent increase in 2004.[7] The largest increases occurred in Asia.[8] China's emissions rose by 9.1 percent in 2005 and experts predict that before 2010 China will emit more carbon from fossil fuel use than the United States does.[9]

In early 2007, the Intergovernmental Panel on Climate Change released its strongest statement yet linking rising CO_2 emissions and increasing global temperatures.[10] Some 2,500 experts concluded with at least 90 percent certainty that the observed warming over the last 50 years has been caused by human activities and that discernible human influences are now apparent in changed precipitation and storm intensity and in other instances of extreme weather worldwide.[11] Heatwaves, floods, and droughts could cause hunger for millions of people and water shortages for billions, with the world's poor hit hardest.[12]

The average global temperature in 2006 was 14.54 degrees Celsius—the fifth warmest year on record, according to NASA's Goddard Institute of Space Studies.[13] (See Figure 3.) Temperatures far above normal were recorded around the globe—from Australia and China to the United Kingdom.[14] Over the past century, average global temperatures have risen nearly 0.06 degrees Celsius a decade, but the rate of increase has tripled since 1976.[15] Eight of the last 10 years rank among the 12 warmest on record.[16]

The climate is warming most rapidly at the poles.[17] Over the past century, Arctic temperatures rose at almost twice the global average rate.[18] For the first time, Inuits now use air conditioners as Arctic summers grow longer and warmer.[19] Nearly 9 percent of the September sea ice in the northern hemisphere is being lost each decade.[20] One model projects that Arctic summers could be ice-free by 2040.[21] In late 2006, the U.S. Interior Department proposed adding polar bears to the list of threatened species as accelerating ice loss threatens their habitat.[22]

A 2006 report compiled for the U.K. government estimated that under business as usual the economic costs of climate change could equal the loss of 5–20 percent of gross world product each year, whereas the cost of efforts to avoid the worst impacts can be limited to about 1 percent of that figure.[23] In early 2007, U.N. Secretary-General Ban Ki-moon warned that upheavals resulting from climate change impacts "from droughts to inundated coastal areas and loss of arable land are likely to become a major driver of war and conflict."[24]

As economic and security concerns intensified in 2006, the general public, businesses, and politicians stepped up their responses. The European Union (EU) carbon market—the world's largest—traded an estimated 1 billion tons of CO_2 emissions, worth more than $19 billion.[25] Carbon prices fell sharply after the release of EU emissions data in May but soon rebounded.[26] In the first nine months of 2006, the global carbon market exceeded $21 billion, more than double the $10 billion traded in 2005, and included countries not bound by the Kyoto Protocol, such as China and India.[27]

In March 2007, EU members agreed to reduce emissions 20 percent below 1990 levels by 2020.[28] At least 12 states in the United States have set emissions targets, and U.S. institutional investors joined 10 leading corporations in calling for a national policy to reduce U.S. emissions.[29]

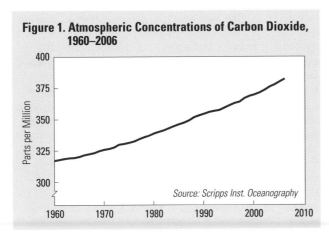

Figure 1. Atmospheric Concentrations of Carbon Dioxide, 1960–2006

Source: Scripps Inst. Oceanography

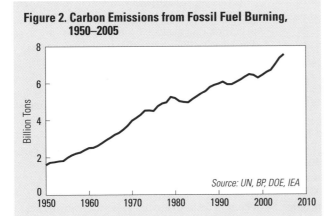

Figure 2. Carbon Emissions from Fossil Fuel Burning, 1950–2005

Source: UN, BP, DOE, IEA

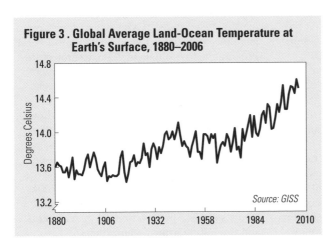

Figure 3. Global Average Land-Ocean Temperature at Earth's Surface, 1880–2006

Source: GISS

Global Average Temperature and Carbon Emissions from Fossil Fuel Burning, 1950–2006, and Atmospheric Concentrations of Carbon Dioxide, 1960–2006

Year	Carbon Dioxide	Emissions	Temperature
	(parts per mill. by vol.)	(bill. tons of carbon)	(degrees Celsius)
1950	n.a.	1.61	13.85
1955	n.a.	2.01	13.90
1960	316.91	2.53	13.99
1965	320.03	3.09	13.89
1970	325.68	4.00	14.03
1975	331.15	4.52	13.95
1980	338.68	5.21	14.18
1985	345.90	5.30	14.06
1986	347.15	5.46	14.13
1987	348.93	5.59	14.27
1988	351.48	5.81	14.31
1989	352.91	5.92	14.19
1990	354.19	5.99	14.38
1991	355.59	6.09	14.35
1992	356.37	5.95	14.13
1993	357.04	5.95	14.14
1994	358.89	6.08	14.24
1995	360.88	6.21	14.38
1996	362.64	6.36	14.30
1997	363.76	6.49	14.40
1998	366.63	6.45	14.57
1999	368.31	6.30	14.33
2000	369.48	6.45	14.33
2001	371.02	6.61	14.48
2002	373.10	6.72	14.56
2003	375.64	7.03	14.55
2004	377.38	7.36	14.49
2005	379.66	7.56	14.63
2006 (prel)	381.84	n.a.	14.54

Source: GISS, BP, IEA, CDIAC, DOE, and Scripps Inst. of Oceanography.

Weather-related Disasters Climb

Zoë Chafe

In 2006, the world experienced more weather-related disasters than in any of the previous three years, according to both Munich Reinsurance Company (Munich Re) and the Center for Research on the Epidemiology of Disasters (CRED).[1] Weather-related disasters include those caused by heat waves or cold snaps, floods, landslides, avalanches, wildfires, hurricanes, cyclones, typhoons, tornadoes, or winter storms.

The economic losses associated with these disasters fell sharply, however, from $219.6 billion in 2005 to $44.5 billion in 2006, according to Munich Re.[2] (See Figure 1.) A relatively quiet Atlantic hurricane season contributed to this dramatic 80-percent decrease in recorded losses.[3] With only 9 named storms, down from 27 in the 2005 season, much of the infrastructure-heavy U.S. coastline that suffered extreme damage in 2005 was spared this year.[4]

CRED recorded a total of 16,193 deaths due to weather-related disasters in 2006, up 24 percent from 12,081 in 2005.[5] Floods were responsible for the greatest number of these deaths in 2006, and they affected 87 countries.[6] (See Figure 2.) The Horn of Africa was particularly hard-hit, experiencing some of the worst monsoon flooding ever recorded.[7] In August, Typhoon Saomai became the strongest storm to make landfall over China in 50 years, destroying 50,000 homes and forcing more than a million people to evacuate.[8]

LINKS pp. 42, 52, 54

Millions of people survive disasters each year, but they continue to suffer long after the flood waters have receded or the storm clouds have disappeared. Between 2002 and 2006, some 827 million people worldwide were affected by weather-related disasters; in 2006 alone, nearly 99 million were affected.[9] (See Figure 3.) This includes 29,400 who were injured and 5.4 million people who became homeless as a result of a disaster.[10]

While weather-related disasters often capture the media spotlight because of their quick onset and dramatic impacts, attention may shift away long before the suffering ends and real recovery begins. This leaves survivors with little support to cope with "secondary" disasters that follow:

sexual harassment in camps, domestic violence, trafficking of children and child labor, poor resettlement plans, and ongoing disabilities.[11]

Weather-related disasters are often perceived as natural events, but many human actions have a hand in their creation. Climate change is warming sea temperatures, which can lead to stronger hurricanes.[12] Sea level rise threatens low-lying areas, especially during storms. Damage to mangrove forests and coral reefs weakens natural storm defenses.[13] And with more people forced to live in undesirable, riskier areas, the potential for disaster is ever higher. Of the 33 cities projected to have at least 8 million residents each by 2015, some 21 are coastal cities that will have to contend with sea level rise.[14]

Cities are particularly vulnerable to weather-related disasters because of their dense infrastructure. But the true economic toll from disasters is difficult to estimate, because most people the world over do not have insurance policies: only 1–3 percent of households and businesses in low-income and middle-income countries are insured against disasters, compared with 30 percent in high-income countries.[15] Only 2 percent of natural disaster losses are covered by insurance in developing countries, while half of such costs are covered in the United States.[16]

U.N. Special Advisor Jeffrey Sachs has recommended that countries secure insurance against frequent natural disasters rather than rely on international aid appeals that are often inadequately funded.[17] The first such national insurance policy was issued by the World Food Programme to Ethiopia in 2006, to protect the residents of that drought-stricken country.[18] If rainfall levels fail to reach an agreed level, farmers will be eligible to receive payouts.[19]

Still struggling with how to mobilize disaster aid as quickly as possible, the United Nations created a new instrument, the Central Emergency Response Fund, to get money and supplies to affected areas within 72 hours of a disaster.[20] Within one year of its March 2006 launch, the fund had received payments and pledges of $343 million from 51 governments and three supporting organizations.[21]

Figure 1. Economic Losses from Weather-Related Disasters, 1980–2006

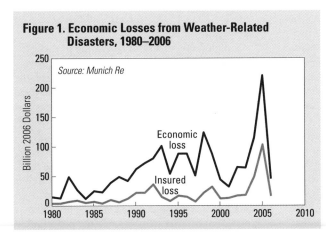

Source: Munich Re

Economic loss

Insured loss

Figure 2. Deaths from Weather-Related Disasters, 2006

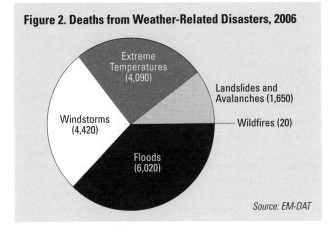

Extreme Temperatures (4,090)

Landslides and Avalanches (1,650)

Wildfires (20)

Windstorms (4,420)

Floods (6,020)

Source: EM-DAT

Figure 3. Number of People Affected by Weather-Related Disasters, 1982–2006

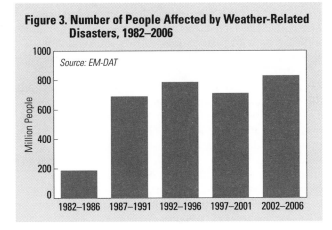

Source: EM-DAT

Economic and Insured Losses from Weather-Related Disasters, 1980–2006

Year	Economic Loss	Insured Loss
	(billion 2006 dollars)	
1980	15	3
1981	13	3
1982	49	6
1983	27	8
1984	12	4
1985	25	6
1986	23	3
1987	39	9
1988	49	5
1989	42	11
1990	62	21
1991	72	21
1992	80	35
1993	101	14
1994	54	7
1995	88	16
1996	88	14
1997	51	6
1998	124	21
1999	88	31
2000	44	11
2001	32	12
2002	65	16
2003	64	17
2004	114	48
2005	220	102
2006 (prel)	45	15

Source: Munich Re.

Ozone Layer Stabilizing But Not Recovered
Alana Herro

After experiencing severe losses between 1979 and 1996, Earth's ozone layer has ceased its precipitous decline, according to scientists with the U.S. National Oceanic and Atmospheric Administration.[1] The average amount of ozone in the stratosphere in 2002–05 was similar to the average measured in 1998–2001, although it was still 3.5 percent below 1964–80 averages.[2] (See Figure 1).

Meanwhile, at its annual peak, the "hole" in the ozone layer above Antarctica grew to 27.5 million square kilometers in 2006—close to the 28.7 million square kilometers reached in 2000.[3] (See Figure 2.) Severe ozone losses are expected there for at least two more decades.[4]

The ozone layer protects Earth from harmful ultraviolet (UV) radiation by absorbing many of the sun's UV rays. But the release into the atmosphere of certain chemicals, such as chlorofluorocarbons (CFCs) and methyl bromide, disrupts the ozone creation cycle, thinning this delicate shield. CFCs have been widely used for refrigeration purposes, aerosol propellants, and blowing agents.

In humans, high levels of UV radiation can cause sunburn and malignant melanoma, lesions and cataracts, and suppression of the immune system; in plants, they can cause DNA damage.[5] When the Antarctic ozone hole widens, people in southern Chile and Argentina are advised to avoid direct sunlight to minimize their health risks.[6]

Much of the success in stabilizing atmospheric ozone levels can be attributed to the Montreal Protocol on Substances That Deplete the Ozone Layer, a treaty adopted in 1987 to reduce the release of ozone-depleting substances (ODS). As a result of scheduled ODS phaseouts in industrial and developing countries, CFC use decreased 96 percent between 1986 and 2005, to 41,200 tons, while methyl bromide use dropped to some 12,500 tons from 37,000 tons in 1995.[7] (See Figure 3.) ODS persist in the stratosphere for many years, however, so decreased use does not immediately mean decreased accumulation. Meanwhile, use of hydrochlorofluorocarbons, a less-damaging CFC substitute that still contributes to some ozone loss, increased steadily—from less than 15,000 tons in 1992 to nearly 32,000 tons in 2005.[8]

Roughly 90 percent of the ozone in the atmosphere is found in the stratosphere, from 10–16 to 50 kilometers above Earth's surface; the rest occurs in the troposphere (from the surface to 10–16 kilometers above).[9] By 2005, total ODS levels in the troposphere had dropped 8–9 percent from their peak in 1992–94; though stratospheric ODS levels peaked in the late 1990s, reductions there are somewhat less because it takes a few years for near-surface trends to be reflected.[10] The stabilization of the ozone layer has stopped the rise in surface UV radiation in unpolluted areas outside the poles and in some areas led to a slight decline in radiation.[11]

Production and use of harmful ODS has not ended completely, however. Exemptions for some ODS, such as methyl bromide for agricultural purposes, are slowing progress.[12] Other challenges include the ongoing illegal trade in CFCs, growing legal production of ODS in developing countries, and the continued use of older refrigerators and other products that contain the chemicals.[13]

The consensus of most researchers is that ozone concentrations over Earth's non-polar regions will return to pre-1980 levels between 2040 and 2050.[14] Ozone concentrations over the Arctic are expected to reach pre-1980 levels at the same time or earlier, while those over the Antarctic are unlikely to do so until 2060–75 (and that is assuming continuing phaseout of ODS).[15] The ozone hole is expected to remain large for at least a decade or so and will continue to fluctuate with meteorological conditions (it is larger in colder winters, for instance).[16]

Cyclic changes in UV radiation emitted by the sun affect ozone levels, since radiation initiates stratospheric ozone formation.[17] A typical solar cycle can contribute to a 1–2 percent variation in total ozone levels.[18] Volcanic eruptions deplete the ozone layer as well, by emitting large amounts of sulfur dioxide, which convert to aerosols that aid chlorine destruction of ozone.[19] Neither of these factors, however, plays as large a role in ozone stability as the release of ODS does.[20]

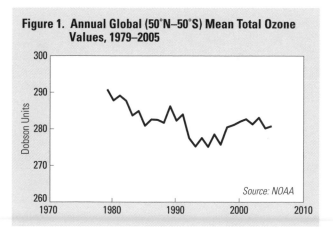

Figure 1. Annual Global (50°N–50°S) Mean Total Ozone Values, 1979–2005

Source: NOAA

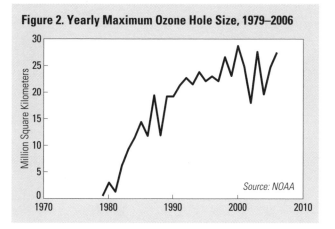

Figure 2. Yearly Maximum Ozone Hole Size, 1979–2006

Source: NOAA

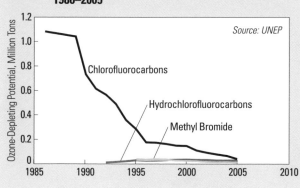

Figure 3. Consumption of Ozone-Depleting Substances, 1986–2005

Source: UNEP

Chlorofluorocarbons

Hydrochlorofluorocarbons

Methyl Bromide

Annual Global (50°N–50°S) Mean Total Ozone Values, 1979–2005	
	(Dobson Units)
1979	290.8
1980	287.8
1981	289.1
1982	287.7
1983	283.7
1984	284.9
1985	280.9
1986	282.6
1987	282.5
1988	281.7
1989	286.2
1990	282.3
1991	284.0
1992	277.5
1993	275.2
1994	277.5
1995	275.1
1996	278.5
1997	275.7
1998	280.5
1999	281.1
2000	282.0
2001	282.7
2002	281.3
2003	283.1
2004	280.2
2005	280.8

Source: NOAA.

Social and Economic Trends

An abandoned bank building serves as home to more than 500 people in Monrovia, Liberia.

▶ Population Rise Slows But Continues

▶ World Is Soon Half Urban

▶ Economy and Strain on Environment Both Grow

▶ Steel Production Soars

▶ Aluminum Production Continues Upward

▶ Gold Mining Output Drops Slightly

▶ Roundwood Production Up

For data and analysis on social and economic trends, including advertising, world trade, and foreign direct investment, go to www.worldwatch.org/vsonline.

Population Rise Slows But Continues

Danielle Nierenberg

The world's population grew to just over 6.5 billion in 2006, a slight increase over the previous year.[1] (See Figure 1.) The population growth rate has declined from a high of 2.1 percent in 1970 to nearly 1.2 percent.[2] (See Figure 2.) But 1.2 percent of 6.5 billion still means some 70 million people added to the world's population each year.[3] (See Figure 3.) U.N. experts project world population will reach 8 billion by 2025 and perhaps 9.1 billion by 2050.[4]

The vast majority of population growth—95 percent—is occurring in developing countries.[5] Africa has the highest growth rate of any region: 2.4 percent annually.[6] Its population is expected to more than double by 2050, to 2.3 billion.[7] While nations such as Ghana and Kenya have experienced substantial declines in the number of births in recent years, women there still bear more than four children in their lifetimes.[8] And in some African nations, women have six children, on average, within their lifetimes.[9]

China, India, and the United States have the largest populations.[10] India's population of 1.1 billion is expected to grow to almost 1.6 billion by 2050. In contrast, as a result of controversial government policies to control population, China's population of just over 1.3 billion today is likely to reach only 1.4 billion by 2050.[11] Today these two nations account for 37 percent of the world's population.[12]

In 2006, the U.S. population reached 300 million people.[13] Populations in many other industrial nations are shrinking—Japan has a fertility rate under 1 percent and Russia is losing more than a million people a year.[14] Yet the United States has a growth rate of 2.8 percent.[15] The environmental and social impacts of this are considerable: People in the United States and Europe on average have a far greater ecological footprint than people in the developing world.[16] They use nearly twice as much fresh water, for example, and more than twice the cropland as people in low-income countries, and they produce 17 times as many carbon emissions.[17]

In contrast to the situation in developing countries, many European nations are worried about falling birth rates and the graying of their populations. In the Czech Republic and the Ukraine, women have on average fewer than 1.2 children, while in Denmark, France, and Norway the figure is about 1.8 and the average age is increasing.[18] Concerns about low fertility rates have prompted some nations to offer incentives—including improved access to child care and paid parental leave—to encourage couples to have more than one child.[19] In France, for example, the government provides free preschool for children.[20]

At the same time, there are more young people on Earth than ever before, creating "youth bulges" in the developing world, where fertility rates are the highest. In more than 100 nations, people aged 15–29 account for nearly half of all adults, and there are concerns that greater numbers of uneducated and poor youth could present a potential security threat, in addition to straining schools and job markets.[21]

Often the nations with the highest fertility and growth rates also have the least access to clean water and adequate, safe sanitation services. In Africa and Asia, 35–50 percent of the population lacks clean drinking water, while 45–60 percent lacks sanitation.[22] At least 1.6 million people die each year of diseases directly related to dirty drinking water and inadequate sanitation.[23]

Lack of access to reproductive health services and family planning methods continues to prevent millions of families from planning and spacing births. An estimated 350 million couples do not have access to contraceptives, and almost 140 million women want to delay their next birth or avoid another pregnancy but are not using any form of birth control.[24]

Complications from pregnancy and childbirth continue to be one of the leading causes of illness and death for women in the developing world. Every year, 8 million women suffer life-threatening complications from pregnancy, and at least 500,000 women die from such complications or during childbirth.[25] Increasing access to safe, affordable, and reliable reproductive health care will help improve the lives of women, men, and children.

LINKS pp. 52, 108

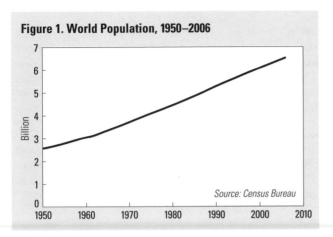

Figure 1. World Population, 1950–2006

Source: Census Bureau

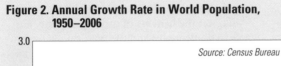

Figure 2. Annual Growth Rate in World Population, 1950–2006

Source: Census Bureau

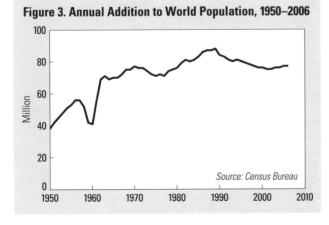

Figure 3. Annual Addition to World Population, 1950–2006

Source: Census Bureau

World Population, Total and Annual Addition, 1950–2006

Year	Total	Annual Addition
	(billion)	(million)
1950	2.56	38
1955	2.78	53
1960	3.04	41
1965	3.35	70
1970	3.71	77
1975	4.08	71
1976	4.15	72
1977	4.23	71
1978	4.30	74
1979	4.37	75
1980	4.45	76
1981	4.52	79
1982	4.60	81
1983	4.68	80
1984	4.76	81
1985	4.84	83
1986	4.93	86
1987	5.01	87
1988	5.10	87
1989	5.19	88
1990	5.27	84
1991	5.36	83
1992	5.44	81
1993	5.52	80
1994	5.60	81
1995	5.68	80
1996	5.76	79
1997	5.84	78
1998	5.92	77
1999	6.00	76
2000	6.07	76
2001	6.15	75
2002	6.22	75
2003	6.30	76
2004	6.37	76
2005	6.45	77
2006 (prel)	6.53	77

Source: U.S. Bureau of the Census.

The United Nations projects that sometime in 2008 more people will live in cities than in rural areas.[1] Over the past half-century, the world's urban population has increased nearly fourfold, from 732 million in 1950 to 3.15 billion in 2005.[2] (See Figure 1.) People living in cities accounted for 49 percent of the total population of 6.46 billion in 2005.[3]

The bulk of future population increase— 88 percent of the growth from 2000 to 2030— is projected to occur in cities of the developing world.[4] Asia and Africa, the most rural continents today, are set to double their urban populations to some 3.4 billion by 2030.[5]

Urbanization has slowed considerably in North America and Europe, where by 1950 more than half the population already lived in cities.[6] Latin America, at 77 percent urban, has also gone through this demographic transition.[7] Growth in that region's "megacities"—urban agglomerations with more than 10 million inhabitants—has slowed, although large slum populations continue to grow, thanks to the world's highest levels of economic and social inequality.[8]

LINKS p. 50

Africa, currently only 38 percent urban, already has nearly 350 million city dwellers— more than the populations of Canada and the United States combined.[9] (See Figure 2.) Urbanization there is more recent and more rapid because of higher population growth, rural poverty, and wars that drive people into cities.[10] Lack of infrastructure for the poor, followed by rapid urban growth, has produced large slum populations at high risk of disease and environmental hazards like flooding.[11] Worldwide, roughly 1 billion urban dwellers live in slums, defined as areas where people live without one or more of life's basic necessities: clean water, sanitation, sufficient living space, durable housing, or secure tenure.[12]

Asia, the world's most populous region, is roughly 40 percent urban.[13] Pacific Asia—the coastal region from Japan to Southeast Asia—has undergone a remarkable economic transformation over the past generation, and China is now the site of 16 of the world's 20 most polluted cities.[14] In western China, South Asia, and interior Asia, urbanization is also rapid, but economic growth has been slower, and poverty burdens nearly a third of India's urban population.[15]

Since 1975, more than 200 urban agglomerations in the developing world have grown past 1 million inhabitants, so local governments are facing greater sanitation, housing, transportation, water, energy, and health care needs.[16] By 2005, 15 of these were megacities (see Figure 3), although these areas account for only about 9 percent of the total urban population.[17] Just over half of the world's city dwellers live in settlements with fewer than 500,000 inhabitants.[18]

More than half of the rise in urban population is caused by natural increase.[19] But migration is also a leading factor, as economic opportunities and improvements in sanitation and clean water have made city life more desirable. Yet the benefits of urban prosperity are shared unequally, and the poor public health conditions of slums still sicken and kill on a large scale.[20]

The environmental challenges that cities face vary with the level of economic activity.[21] The poorest cities and their slums typically have the worst local hazards, such as diseases spread by dirty water and lack of toilets.[22] As a city industrializes, problems at the metropolitan scale, such as air pollution from industry and traffic, tend to worsen first and then improve as economic growth allows for cleaner technologies.[23] But a city's burden on the global environment often increases with economic growth as residents buy more cars, bigger houses, and other consumer goods.[24]

Yet the economies of scale possible with high-density settlement provide a crucial opportunity to create living patterns in harmony with nature's rhythms. Urban planners are beginning to embrace the concept of "circular metabolism," in which much of the waste from the water, food, fuels, and materials that course through cities is reused or recycled.[25] Architects are beginning to apply this idea to buildings: the 15-story IBM headquarters in Kuala Lumpur, Malaysia, for example, uses plantings on its exterior to capture water that would otherwise be wasted.[26]

Figure 1. World Urban Population, 1950–2005

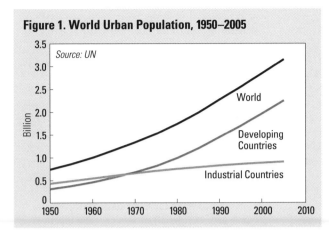

Source: UN

World Urban Population, 1950–2005

Year	World	Industrial Countries	Developing Countries
		(million)	
1950	732	423	309
1955	850	477	374
1960	992	535	458
1965	1,158	595	563
1970	1,329	650	678
1975	1,516	701	815
1980	1,736	744	992
1985	1,984	780	1,204
1990	2,271	818	1,453
1995	2,551	849	1,702
2000	2,845	874	1,971
2005	3,150	898	2,252

Source: UN Population Division.

Figure 2. Urban Population by Region, 1950, 1990, and 2005

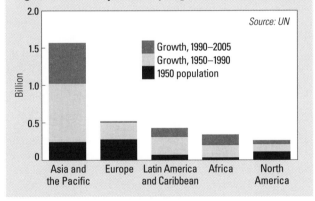

Source: UN

Figure 3. Population of 14 Largest Cities, 1950, 1990, and 2005

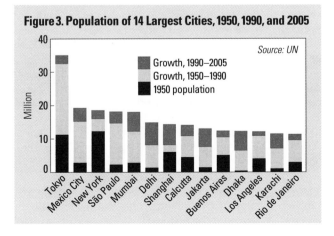

Source: UN

Economy and Strain on Environment Both Grow   Erik Assadourian

In 2006, the gross world product (GWP)—the aggregated total of all finished goods and services produced worldwide—increased 3.9 percent to $65.1 trillion (in 2006 dollars).[1] (See Figure 1.) This estimate reflects real purchasing power in countries (that is, in purchasing power parity or PPP terms). The market exchange rate GWP, which is based on actual monetary terms, reached $47.8 trillion in 2006, an increase of 4.7 percent.[2] Growth of GWP (PPP) in 2006 was slightly less than the 4.0-percent increase in 2005 but about 0.4 percent higher than the average growth seen since 1971.[3]

China accounted for over one third of the $2.5 trillion in growth in 2006.[4] The Chinese economy was once again the fastest growing in the world, with its gross domestic product (GDP) jumping 8.8 percent, driven mainly by high levels of investment and exports.[5] Yet analysts increasingly question whether China can sustain this growth, as the benefits have been distributed unequally and have also created significant environmental problems.[6] In 2006, accidents triggering pollution that the Chinese government considered "serious" occurred almost every other day on average.[7]

LINKS pp. 44, 112, 114

Sub-Saharan Africa, the Middle East, and Russia and the former states of the Soviet Union also grew at a fast clip, of 5.1 percent, 4.2 percent, and 5.1 percent respectively.[8] This growth primarily stemmed from strong net exports of commodities, particularly oil and natural gas and, in sub-Saharan Africa, metals.[9]

The U.S. economy, accounting for 20 percent of GWP, grew 2.7 percent in 2006.[10] The United States thrived in the first quarter, but high fuel prices, sluggish job growth, and a weakening housing market slowed economic expansion later.[11] With continued cooling of the housing market, consumer demand and economic growth are expected to slow further in 2007.[12]

The European Union also accounted for 20 percent of GWP in 2006.[13] Its economy grew 1.5 percent, primarily driven by domestic spending and investment.[14] Job growth in the United Kingdom and consumer demand in Germany contributed to this increase.[15] Japan grew at 1.3 percent in 2006, with strong domestic demand offset by a reduction in public investment and net exports.[16]

Per capita GWP also increased in 2006, to $9,975.[17] This is a growth of 2.7 percent—less than total GWP growth because world population increased by 77 million people.[18] Yet GWP per capita does not reflect the vast disparity in GDP per person—even when these figures are in purchasing power parity terms. In the United States GDP is $43,356 per person and in Japan it is $31,924, for example, while in China the figure is $8,005 and in India it is $3,546.[19]

GDP is a poor measure of economic progress, as it counts all monetary expenditures as positive—whether the money is spent on useful goods, such as food or durables, or on mitigating social ills that could have been prevented. The U.S. nongovernmental organization Redefining Progress designed the Genuine Progress Indicator (GPI), a measure that better analyzes economic progress by subtracting out pollution and resource degradation, crime, and other economic ills while adding in unmeasured benefits like volunteer work and parenting.[20] While U.S. GDP per capita has nearly doubled since 1970, the GPI grew just 15 percent.[21] (See Figure 2.)

Clearly, economic priorities must change, as over 60 percent of ecosystem services are being degraded or used unsustainably.[22] The "ecological footprint" of global society—a measurement that calculates the amount of land and sea area needed to produce resources, absorb wastes, and provide space for infrastructure, such as roads and buildings—is also increasing each year, with a jump of 2.5 percent in 2003.[23] (See Figure 3.)

This most recent measurement shows that humans currently use the resources of 1.25 Earths and are thus depleting the ecological capital on which future populations will depend.[24] As economic growth accelerates in both high-income and low-income countries, so does the depletion of ecological capital. Indeed, at the current consumption levels of high-income countries, the world could only sustainably support 1.75 billion people, not the 6.5 billion living on Earth today.[25]

Figure 1. Gross World Product, 1970–2006

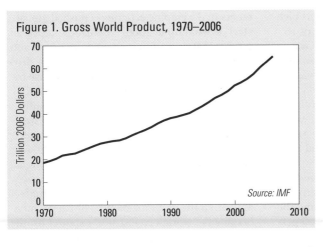

Source: IMF

Figure 2. GDP and GPI Per Person, United States, 1950–2004

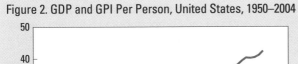
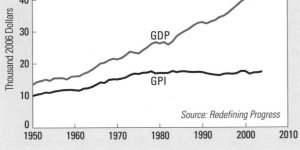

GDP

GPI

Source: Redefining Progress

Figure 3. Humanity's Ecological Footprint, 1961–2003

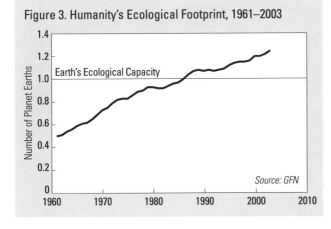

Earth's Ecological Capacity

Source: GFN

Gross World Product, 1970–2006		
Year	Total	Per Capita
	(trill. 2006 dollars)	(2006 dollars)
1970	18.6	5,006
1971	19.4	5,124
1972	20.4	5,281
1973	21.8	5,530
1974	22.3	5,568
1975	22.7	5,561
1976	23.8	5,731
1977	24.9	5,881
1978	26.0	6,049
1979	27.0	6,167
1980	27.6	6,200
1981	28.1	6,223
1982	28.4	6,176
1983	29.3	6,252
1984	30.6	6,434
1985	31.8	6,561
1986	32.9	6,683
1987	34.2	6,821
1988	35.8	7,016
1989	37.1	7,151
1990	38.1	7,225
1991	38.7	7,220
1992	39.5	7,255
1993	40.3	7,306
1994	41.8	7,458
1995	43.3	7,614
1996	45.0	7,814
1997	46.9	8,034
1998	48.2	8,141
1999	49.9	8,327
2000	52.3	8,611
2001	53.6	8,719
2002	55.2	8,873
2003	57.4	9,110
2004	60.3	9,452
2005	62.7	9,712
2006 (prel)	65.1	9,975

Source: IMF.

Steel Production Soars

Yingling Liu

Global crude steel production hit a record high of 1.24 billion tons in 2006, an increase of 10 percent over 2005.[1] (See Figure 1.) This was the third consecutive year in which crude steel output exceeded 1 billion tons.[2] China was by far the leading producer, with 419 million tons output in 2006—just above one third of the global total.[3] The other major producers were Japan (116 million tons) and the United States (99 million tons), followed by Russia and South Korea.[4] (See Figure 2.)

The past decade has been the most productive in the history of the steel industry, driven mainly by remarkable growth in China and the Asia region. Global output in 2006 was 65 percent above the figure a decade earlier.[5] China became the largest national producer in 1996, and 10 years later output there was a startling 314 percent higher.[6] The Asia region accounted for 38 percent of all crude steel produced in 1996; by 2006, the share rose to 54 percent.[7]

LINKS p. 58

Consolidation in the steel industry worldwide has accelerated as producers look to integrate horizontally with other mills and vertically with raw material suppliers and steel distributors to secure their futures.[8] In 2005 the top 15 steel producers accounted for one third of world production, compared with just over one fourth in 1995.[9] (See Figure 3.)

The recent race toward consolidation has been highlighted by a few major takeovers. In June 2006 Mittal Steel took over the Pan-European Arcelor and became the largest steelmaker in the world.[10] The new firm, Arcelor-Mittal, has more than 100 million tons of annual capacity—enough for twice as many automobiles as are made in the world every year and three times the capacity of its nearest rival, Nippon Steel.[11]

The second major merger took place in early 2007, when an Indian conglomerate—Tata Steel—acquired the Anglo-Dutch steel firm Corus and created the world's fifth biggest steel producer.[12] Similar takeovers and mergers also happened in the United States, Europe, Russia, China, East Asia, and Australia.[13]

The rebounding world economy combined with buoyant infrastructure and other investments in developing economies pushed global steel demand up in 2006.[14] Demand jumped an estimated 9 percent in the year, with China and, more generally, Asia again being the major driving forces.[15] Demand for steel in China rose by 15 percent in 2006, accounting for one third of the global total.[16] Increased spending on infrastructure and construction in India drove steel use there up by 10 percent.[17] Demand for steel also rose considerably in the European Union, North America, East Asia, and Russia, though at more moderate rates.[18]

Rising global demand has stimulated trade as well. In the first nine months of 2006 China became the world's largest steel exporter for the first time, surpassing Japan, Russia, and the European Union.[19] China's net exports for the year reached 24.5 million tons.[20] North America and the European Union remained the key steel-importing regions, with each estimated to bring in around 40 million tons a year.[21]

Recycled iron and steel scrap is a vital raw material, and the rate of recycling has risen remarkably in industrial countries. The recycling rate for steel reached 76 percent in the United States in 2005, the highest ever recorded there.[22] In 2006 an estimated 55 million tons of steel were recycled in the United States.[23] The latest available data show that the U.S. recycling rate in 2005 for automobiles—the primary source of old steel scrap—was 102 percent, indicating that more steel was reclaimed from automobiles than was used to manufacture new vehicles.[24]

The U.S. recycling rates for appliances and steel cans in 2006 were 90 percent and 63 percent, respectively.[25] Worldwide, more than 5 million tons of steel cans were recycled in 2005, an average recycling rate for steel packaging of 65 percent.[26] This figure is 7.4 percent higher than in 2001, signaling a continuous increase over the years.[27]

Figure 1. World Steel Production, 1950–2006

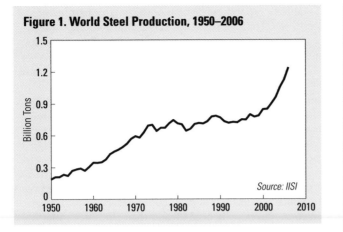

Source: IISI

Figure 2. Top Five Steel-Producing Countries, 1994–2006

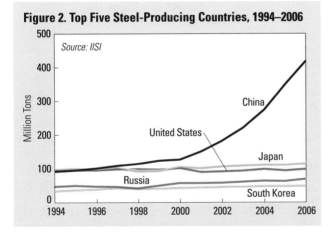

Source: IISI

Figure 3. Top 15 Steel-Producing Companies, 2005

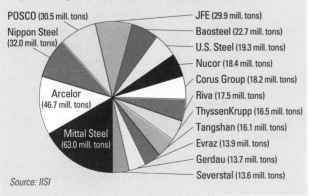

POSCO (30.5 mill. tons)
Nippon Steel (32.0 mill. tons)
Arcelor (46.7 mill. tons)
Mittal Steel (63.0 mill. tons)

JFE (29.9 mill. tons)
Baosteel (22.7 mill. tons)
U.S. Steel (19.3 mill. tons)
Nucor (18.4 mill. tons)
Corus Group (18.2 mill. tons)
Riva (17.5 mill. tons)
ThyssenKrupp (16.5 mill. tons)
Tangshan (16.1 mill. tons)
Evraz (13.9 mill. tons)
Gerdau (13.7 mill. tons)
Severstal (13.6 mill. tons)

Source: IISI

World Steel Production, 1950–2006

Year	Production (million tons)
1950	190
1955	271
1960	347
1965	451
1970	595
1975	644
1976	675
1977	675
1978	717
1979	747
1980	716
1981	707
1982	645
1983	664
1984	710
1985	719
1986	714
1987	736
1988	780
1989	786
1990	771
1991	734
1992	720
1993	728
1994	725
1995	752
1996	750
1997	799
1998	777
1999	788
2000	847
2001	850
2002	904
2003	962
2004	1,057
2005	1,129
2006 (prel)	1,240

Source: IISI.

Aluminum Production Continues Upward

Gary Gardner

Global production of primary aluminum—aluminum made from bauxite ore—increased by 4 percent in 2006.[1] (See Figure 1.) That number represents a continuing slowdown in output in recent years from the high rates of 2002–05, when production increases averaged 6.5 percent annually.[2] Nevertheless, the industry continues to grow globally as demand moves upward and as new production capacity is added.[3] Meanwhile, global secondary (recycled) aluminum production was up in 2004, the latest year for which world data are available.[4]

Aluminum is the world's second most popular metal, after iron. It is used to make transport vehicles from cars to airplanes to ships, in construction, in consumer durables such as appliances, and in packaging.[5] Aluminum is made from bauxite ore, which is found near the earth's surface and which usually requires open-pit mining to be removed.[6] Bauxite is relatively plentiful, with the greatest reserves found in Guinea, Australia, Brazil, Jamaica, and China.[7]

Primary aluminum production is concentrated in relatively few countries. China alone produced 26 percent of the world total in 2006.[8] The top five producers—China, Russia, Canada, the United States, and Australia—accounted for 59 percent of world output that year.[9] (See Figure 2.) Production is found where energy is cheap because making aluminum uses gargantuan quantities of electricity.[10] Indeed, the world's largest aluminum smelter, now being planned for construction in Dubai, will have its own 2,600-megawatt power plant.[11]

Although growth in primary production has slowed, many of the top producers posted robust gains. China and India reported the greatest increases in aluminum output, at 12 percent and 11 percent respectively.[12] Bahrain, Brazil, and South Africa also posted large gains for the year, at 11, 7, and 5 percent.[13]

Growth globally was slowed by the 7-percent decline in production experienced in the United States, the fourth largest producer.[14] This decline is part of a long-term pattern: primary production in this country has fallen by 57 percent since 1992.[15] U.S. smelters were operating at only 62 percent of capacity, in part because of high prices for energy and alumina.[16] Meanwhile, secondary aluminum production in the United States may also experience a slowdown as automobile manufacturing, a key supplier of scrap, declines there.[17]

Aluminum production accounts for roughly 3 percent of global electricity use.[18] For some countries the share is much higher: in Australia, it devours 10 percent of the country's electricity.[19] The industry has become steadily more energy-efficient in recent years; electricity use per ton of output fell by 11 percent between 1980 and 2006.[20] But increases in output have typically been greater than efficiency gains, sending total electricity use for aluminum higher each year.[21] (See Figure 3.)

Aluminum from scrap (from manufacturing plants as well as aluminum products) reduces this metal's environmental footprint because of its relatively low energy requirements. In the United States in 2006, roughly two thirds of aluminum used in recycling came from manufacturing plants; the remainder came from discarded products.[22] Aluminum recycled from discarded products accounted for the equivalent of about 18 percent of aluminum consumption in the United States in 2006.[23]

The growing practice of making aluminum from scrap rather than from virgin ore will affect the location and economics of production in the future.[24] Remelting (recycling) aluminum uses only 5–10 percent as much energy as making aluminum from ore.[25] And because the energy efficiency of recycling aluminum is expected to increase faster than the efficiency of virgin production in coming years, the cost advantage will likely tilt further in the direction of recycling.[26] This cost advantage, coupled with growing availability of scrap aluminum, are likely to decouple aluminum production from supplies of cheap energy.[27]

Aluminum, if recycled, has a number of environmentally friendly features, including light weight, which means products require less energy to transport, and strength, which means less is needed for a given function.[28] Aluminum can be recycled many times over.

LINKS p. 56

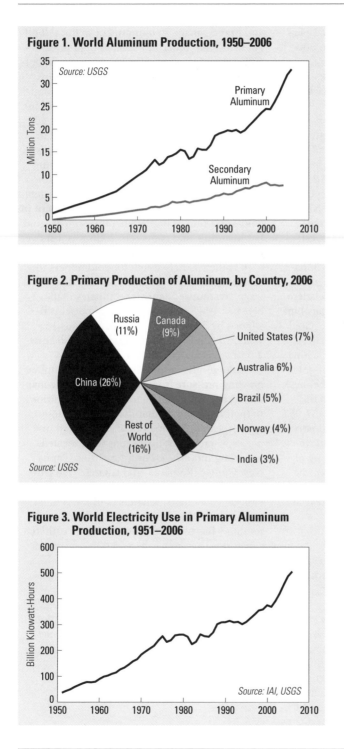

Figure 1. World Aluminum Production, 1950–2006

Source: USGS

Primary Aluminum

Secondary Aluminum

Figure 2. Primary Production of Aluminum, by Country, 2006

Russia (11%)

Canada (9%)

United States (7%)

Australia 6%

Brazil (5%)

China (26%)

Norway (4%)

Rest of World (16%)

India (3%)

Source: USGS

Figure 3. World Electricity Use in Primary Aluminum Production, 1951–2006

Source: IAI, USGS

World Aluminum Production, 1950–2006

Year	Primary	Secondary
	(million tons)	
1950	1.5	0.4
1955	3.1	0.6
1960	4.5	0.9
1965	6.3	1.5
1970	9.7	2.2
1975	12.1	2.8
1976	12.6	3.1
1977	13.8	3.4
1978	14.1	4.0
1979	14.6	3.8
1980	15.4	3.9
1981	15.1	4.1
1982	13.4	3.8
1983	13.9	4.1
1984	15.7	4.2
1985	15.4	4.4
1986	15.4	4.5
1987	16.5	4.8
1988	18.5	5.3
1989	19	5.4
1990	19.3	5.8
1991	19.7	5.6
1992	19.5	5.7
1993	19.8	6.3
1994	19.2	6.6
1995	19.7	7.0
1996	20.7	6.9
1997	21.6	7.4
1998	22.6	7.5
1999	23.6	7.9
2000	24.4	8.2
2001	24.3	7.6
2002	25.9	7.7
2003	27.7	7.5
2004	29.8	7.6
2005	31.9	n.a.
2006 (prel)	33.1	n.a.

Source: USGS.

Gold Mining Output Drops Slightly

Yingling Liu

In 2006, mine production of gold fell by 1 percent to 2,467 tons, in line with output in 2004.[1] (See Figure 1.) The world's top three producers—South Africa, Australia, and the United States—and Indonesia all saw marked losses in production.[2]

The main shaft at South Deep in South Africa was temporarily closed, slashing production there by half.[3] The Batu Hijau Mine in Indonesia suffered production losses due to pit wall stability issues.[4] The operators of the world's two largest gold-producing mines—Freeport-McMoRan at Grasberg in Indonesia and Newmont at Yanacocha in Peru—both reported substantial reductions in output in 2006, estimated at a combined 75 tons for the year.[5]

Some Latin American countries and China achieved slight production increases when a number of new mines got going.[6] The Veladero mine in Argentina and the Amapari and Jacobina mines in Brazil all saw increased output.[7] Gold mining in China increased steadily as well. As the world's fourth-largest gold producer, China produced a record amount of 240 tons of gold in 2006—7 percent more than the year before.[8]

The gold produced in mines added to the existing gold stockpile, bringing this to 157,967 tons at the end of 2006.[9] The jewelry industry accounted for 52 percent of the above-ground gold stocks at the beginning of 2006, central banks held 18 percent in their vaults, private investors hoarded 16 percent, and 12 percent was used for industrial purposes.[10] (See Figure 2.)

LINKS p. 54

Gold prices continued to climb in 2006.[11] (See Figure 3.) Gold at the London price touched $725.75 in mid-May—the highest level in 25 years.[12] The average price throughout the year was $604, up nearly 36 percent from 2005.[13]

Shrinking sales by central banks were partly responsible for the price hike in 2006.[14] Net sales by these banks were estimated to have halved, dropping to 330 tons.[15] The decline was somewhat driven by lower sales from members of the renewed Central Bank Gold Agreement (CBGA-2), an agreement under which 15 of the world's biggest gold holders, including Germany and France, made a commitment to not sell off gold stocks in order to maintain high prices.[16] CBGA members sold only 393 tons of gold out of the possible annual allowance of 500 tons.[17]

The high and volatile prices for gold dampened consumers' enthusiasm. Gold demand for jewelry fabrication slumped by more than 400 tons in 2006. The greatest losses occurred in the price-sensitive regions of India and the Middle East, while Italy and East Asia (excluding China) also saw substantial declines.[18] Chinese retail sales of gold jewelry, in contrast, rose more than 20 percent in 2006.[19] World investment in gold—the sum of implied net investment, gold bar hoarding, and coins—in 2006 was just over 680 tons, down 16 percent.[20]

It is increasingly tough for gold miners to replace gold in the ground with new discoveries.[21] Westhouse Securities estimates that between 1985 and 2003, new gold discoveries slipped by 30 percent from the previous 15 years.[22] Each new ounce discovered also costs 2.6 times as much to locate.[23]

Gold mining corporations have been under growing pressure from the public, jewelry manufacturers, and retailers to pay more attention to their environmental impact. Public protests against leading companies were reported in places where major operations were located, including Indonesia, Peru, Argentina, Papua New Guinea, Romania, and Ghana.[24] No Dirty Gold—an international consumer campaign to educate consumers, retailers, and the general public about the impacts of irresponsible gold mining—has over the past three years gathered signatures from more than 55,000 consumers on a petition urging jewelry retailers to sell environmentally and socially responsible gold.[25] As of early March 2007, 21 leading jewelry retailers—including Cartier, the Zale Corp., Tiffany & Co., and Birks & Mayors—had endorsed No Dirty Gold's Golden Rules, a set of social, environmental, and human rights principles to guide more responsible production.[26]

Figure 1. Global Gold Production, 1950–2006

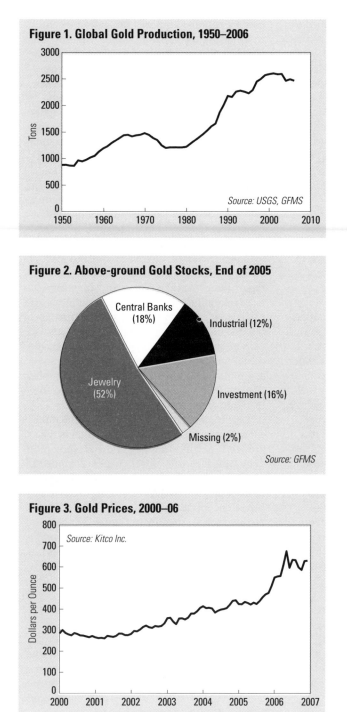

Source: USGS, GFMS

Figure 2. Above-ground Gold Stocks, End of 2005

Central Banks (18%)

Industrial (12%)

Jewelry (52%)

Investment (16%)

Missing (2%)

Source: GFMS

Figure 3. Gold Prices, 2000–06

Source: Kitco Inc.

Global Gold Production, 1950–2006	
Year	Amount
	tons
1950	879
1955	947
1960	1,190
1965	1,440
1970	1,480
1975	1,200
1976	1,210
1977	1,210
1978	1,210
1979	1,210
1980	1,220
1981	1,280
1982	1,340
1983	1,400
1984	1,460
1985	1,530
1986	1,610
1987	1,660
1988	1,870
1989	2,010
1990	2,180
1991	2,160
1992	2,260
1993	2,280
1994	2,260
1995	2,230
1996	2,290
1997	2,450
1998	2,500
1999	2,570
2000	2,590
2001	2,604
2002	2,587
2003	2,593
2004	2,464
2005	2,494
2006 (prel)	2,467
Source: USGS, GFMS.	

Roundwood Production Up

Gary Gardner

Roundwood production worldwide climbed to 3,503 million cubic meters in 2005, the last year with global data.[1] (See Figure 1.) That represents a 2.3-percent increase over 2004, a substantial acceleration of the average 0.64-percent rate of growth of the previous five years.[2] On a per capita basis, however, global production has dropped steadily for more than four decades, from 0.76 cubic meters per person in 1961 to 0.54 cubic meters in 2005, as harvesting and processing technologies have become more efficient and as other materials have replaced wood in some applications.[3]

Roundwood refers to wood that is removed from forests or other areas, whether felled or simply picked up from the forest floor.[4] There are two broad categories: fuelwood (used for heating and cooking) and industrial wood products, such as lumber, wood panels, and wood pulp.

While output of the various kinds of roundwood is split nearly evenly at the global level (51 percent fuelwood and 49 percent industrial), the two products often have different values in wealthy and poor nations.[5] Some three quarters of the world's fuelwood is burned in developing countries, where it accounts for 15 percent of primary energy use.[6] The other one quarter is consumed in industrial countries—in wood stoves, for example—and constitutes only 2 percent of those countries' primary energy supply.[7]

LINKS | p. 54

The top six producers—the United States, India, China, Brazil, Canada, and Russia—accounted for 48 percent of global production in 2005.[8] (See Figure 2.) All showed increases in harvesting over 2004 except China, which sharply curtailed cutting after floods caused by denuded hillsides devastated the country in 1998.[9] Yet China is a major player on the global wood stage: its imports have more than tripled since 1997, making China the world's largest importer of wood and wood products.[10] Demand there has helped fuel the increased output of some major exporters: Chinese imports of Russian logs increased 21-fold between 1997 and 2005, for example.[11]

Overall, the trend in wood harvesting is up in most regions, primarily because of demand from the rapidly expanding economies of countries like China.[12] Rising fossil fuel prices have stimulated demand for wood as a source of heat.[13] Government promotion of renewable energy and climate change policies in many countries, which often steer economic activity away from fossil fuels, is another factor.[14] On the other hand, rising U.S. and European interest rates in 2005–06 helped dampen demand for wood panels and lumber.[15]

Illegal logging is a major obstacle to making forest practices sustainable, because illegally sourced wood and wood products supplied at submarket rates tend to undercut responsibly produced products in world markets.[16] Illegal logging is primarily driven by demand for cheap products in industrial nations. A 2004 report found that the European Union imports nearly 3 billion euros (almost $4 billion) worth of illegal logs—a substantial share of the 10–15 billion euros worth of illegal logging worldwide each year.[17] Illegal logging is also facilitated by illegal products being imported to China from, say, Indonesia and Papua New Guinea and then re-exported to industrial countries, especially the United States and Europe.[18]

One bright spot in the effort to combat illegal and unsustainably produced wood is the growth in certification of wood and wood products. Global certified forested area expanded by 12 percent in 2005, bringing the certified share of the world's forests to 7 percent.[19] About 24 percent of roundwood production comes from certified forests, mainly in Europe and North America.[20] (See Figure 3.) Most of this wood is not labeled, apparently because consumers have yet to demand it.[21]

Most certified forest area is located in industrial countries: North America has 58 percent of the current total and Western Europe has 29 percent.[22] About half of the forested area of Europe and about a third of the area of North America is certified, while nearly all the forests in Austria and Finland are.[23] In absolute terms, Canada has by far the largest certified area—some 121 million hectares.[24] The United States is second, with 35 million hectares.[25]

Figure 1. World Roundwood Production, 1961–2005

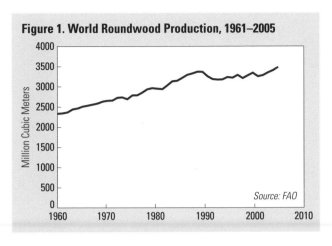

Figure 2. Top Six Roundwood-Producing Countries, 1961–2005

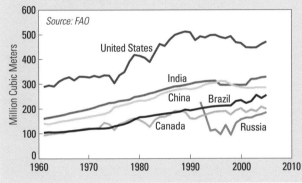
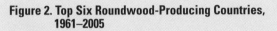

Figure 3. Source of Roundwood from Certified Forests, 2006

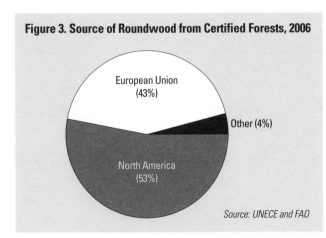

World Roundwood Production, 1961–2005	
Year	Production
	(million cubic meters)
1961	2,342
1965	2,475
1966	2,520
1967	2,543
1968	2,571
1969	2,598
1970	2,644
1971	2,666
1972	2,672
1973	2,740
1974	2,749
1975	2,705
1976	2,799
1977	2,801
1978	2,868
1979	2,950
1980	2,978
1981	2,965
1982	2,954
1983	3,048
1984	3,148
1985	3,162
1986	3,233
1987	3,309
1988	3,343
1989	3,386
1990	3,382
1991	3,272
1992	3,201
1993	3,187
1994	3,193
1995	3,251
1996	3,234
1997	3,305
1998	3,224
1999	3,293
2000	3,358
2001	3,271
2002	3,299
2003	3,368
2004	3,423
2005	3,503

Source: FAO.

Transport and Communications Trends

A family checks their cell phones in La Paz, Bolivia.

▶ Vehicle Production Rises Sharply

▶ Bicycle Production Up Slightly

▶ Air Travel Reaches New Heights

▶ Cell Phones Widely Used, Internet Growth Slows

For data and analysis on transport and communications trends, including car-sharing and passenger rail travel, go to www.worldwatch.org/vsonline.

Vehicle Production Rises Sharply

Michael Renner

According to preliminary figures from London-based Global Insight, global passenger car production grew to 48.6 million units in 2006, a sharp 6 percent increase from 2005.[1] In addition, production of "light trucks" remained unchanged at 18.5 million units, for a combined total of 67.1 million vehicles.[2] (See Figure 1.)

The year 2006 saw some momentous changes in the lineup of top producer countries. Japan, with an output of 10.9 million cars and light trucks, edged out the United States (10.8 million) for the first time.[3] More dramatically, China increased its production by 29 percent to 6.7 million vehicles and overtook Germany (5.7 million) to become the third-largest producer.[4] (See Figure 2.) South Korea and France (3.8 million and 3.2 million, respectively) were next.[5] Spain, Canada, Brazil, and Mexico round out the top 10, but India is poised to break into that league soon.[6]

Vehicle exports—totaling 23.4 million cars and 4.2 million commercial vehicles in 2005—are dominated by just three nations: Japan, France, and Germany together account for 49 percent of the world total.[7]

China's rise represents the most dramatic change in the world auto industry. During the past decade production there more than quintupled, driven by joint ventures with multinationals and the rapid growth of indigenous carmakers.[8] Sales within China surpassed the 3 million mark in 2005, with 8.9 million passenger cars on the country's roads—still a comparatively small fleet.[9] But China is expected to become a major exporter within the next four years.[10] Fuel economy is a key requirement to succeed in the Chinese market and will likely be a strong characteristic of the global market strategy of China's manufacturers.[11]

Worldwide production is concentrated among just a few companies. The top 5—General Motors (GM), Toyota, Ford, Volkswagen, and DaimlerChrysler—manufactured 51 percent of all passenger cars and light trucks in 2005, while the top 10 accounted for 76 percent.[12] (See Figure 3.) A changing of the guard at the top is in the making, however. Toyota passed Ford to become the world's second-largest producer in 2005.[13] Following a $12.7-billion loss for 2006, the worst in its history, Ford will further consolidate rather than grow.[14] Expectations are that Toyota will next—perhaps as soon as 2007—topple GM, the world's largest carmaker since 1931.[15]

Toyota already enjoys a solid lead in hybrid gasoline-electric cars, selling 312,500 hybrids worldwide in 2006 and expecting to sell 430,000 in 2007.[16] However, this and other advances in fuel efficiency technology are being more than offset by carmakers' continued emphasis on adding size and muscle to vehicles. In the United States, the 2006 model year includes the heaviest vehicles in three decades.[17] Although the U.S. government announced new fuel economy standards for light trucks for the model years 2008 to 2011, these are expected to save the equivalent of less than a month's worth of current fuel consumption.[18]

Diesel engine–powered cars typically get 30 percent better mileage than cars with gasoline engines.[19] Consulting firm J.D. Power and Associates projects that global demand for diesel light vehicles will nearly double from 15 million in 2005 to 29 million in 2015.[20] In addition to the traditional stronghold Europe, South Korea and India are key markets for these cars.[21]

The continued expansion of car-centered transportation contributes to climate change. Carmakers selling vehicles in the European Union (EU) agreed in the late 1990s to voluntarily reduce carbon emissions to 140 grams per kilometer by 2008–09—but appear unable to meet this target.[22] New cars made by European companies emitted 161 grams of carbon in 2004, Korean models 168 grams, and Japanese cars 170 grams.[23] EU mandatory limits are in the works that would require a further cut to 130 grams by 2012—though this is criticized by environmentalists as insufficient.[24] In Canada, too, there is growing pressure to introduce binding limits in the face of shortcomings of voluntary measures.[25] And California sued six of the world's largest automakers over climate change, charging that greenhouse gases from their vehicles have caused billions of dollars in damages.[26]

Figure 1. World Passenger Vehicle Production, 1950–2006

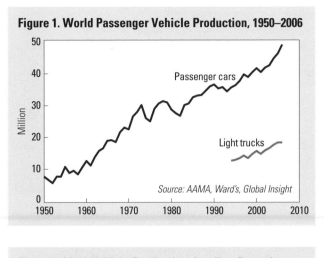

Passenger cars

Light trucks

Source: AAMA, Ward's, Global Insight

Figure 2. Light Vehicle Production. Leading Countries, 1995–2006

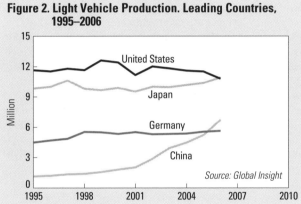

United States

Japan

Germany

China

Source: Global Insight

Figure 3. Market Shares of Leading Automobile Manufacturers, 2005

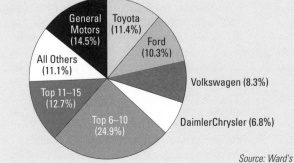

General Motors (14.5%)
Toyota (11.4%)
Ford (10.3%)
All Others (11.1%)
Top 11–15 (12.7%)
Top 6–10 (24.9%)
Volkswagen (8.3%)
DaimlerChrysler (6.8%)

Source: Ward's

World Passenger Vehicle Production, 1950–2006

Year	Passenger Cars	Light Trucks
	(million)	
1950	8.0	
1955	11.0	
1960	12.8	
1965	19.0	
1970	22.5	
1975	25.0	
1976	28.9	
1977	30.5	
1978	31.2	
1979	30.8	
1980	28.6	
1981	27.5	
1982	26.7	
1983	30.0	
1984	30.5	
1985	32.4	
1986	32.9	
1987	33.1	
1988	34.4	
1989	35.7	
1990	36.3	
1991	35.1	
1992	35.5	
1993	34.2	
1994	35.4	12.9
1995	36.1	13.2
1996	37.4	13.7
1997	39.4	14.5
1998	38.6	13.7
1999	40.1	15.0
2000	41.3	15.9
2001	40.1	15.0
2002	41.5	16.1
2003	42.2	16.8
2004	44.4	17.8
2005	45.9	18.5
2006 (prel)	48.6	18.5

Source: American Automobile Manufacturers Association, Ward's, and Global Insight.

Bicycle Production Up Slightly

Gary Gardner

Bicycle production rose to 105 million units globally in 2004 (the last year with global data), a 1.5-percent increase over 2003.[1] (See Figure 1.) The increase is actually a slowdown in production as firms draw down inventories that had grown during two years of rapid growth.

Although bicycles are produced in dozens of countries, the top five producers—China, India, the European Union, Taiwan, and Japan—are responsible for 87 percent of global production.[2] China is in a league of its own, however, with some 58 percent of the global market in 2004.[3] (See Figure 2.)

Preliminary figures suggest that Chinese production increased by 11 percent in 2005.[4] Given the relatively flat trajectories of other major producers, it is very possible that by 2006 China alone accounted for two thirds of global bicycle output.

LINKS p. 52

The Chinese juggernaut faces obstacles, however, especially trade barriers imposed in Europe, Mexico, and Canada. Mexico's 144-percent tariffs, first imposed in 1994 and renewed in 2005, have effectively shut Chinese bikes out of that country.[5] And Europe's 48.5-percent duties on Chinese bicycles, adopted in 2005, were expected to curb Chinese sales in Europe substantially.[6] But China may be adapting to the restrictive European market, in part by setting up production facilities in Eastern Europe.[7] Meanwhile, India is reducing tariffs on bicycles, an encouraging development for Chinese firms, especially in light of projections of a 33-percent increase in demand for bicycles there between 2006 and 2010.[8]

Production of electric bicycles—with electric motors that make longer and hillier rides possible with less exertion—continues to boom. Introduced in the mid-1990s, global sales rose to 12.1 million units by 2005.[9] China accounted for as many as 11 million of these and is expected to remain the strongest engine of demand: Chinese consumers were expected to buy 15 million electric bikes in 2006.[10] Meanwhile, electric models could receive a boost when new standards to be finalized in Europe make components of different producers compatible.[11]

Government support for bicycles—typically in response to concerns about climate change, traffic congestion, energy conservation, and fitness—is increasingly evident. Taiwan announced in 2006 a plan to build 2,600 kilometers of bicycle lanes over the next five years and to adopt bicycle-friendly traffic rules.[12] In London, a congestion tax on autos produced a 50-percent increase in bike trips in the city, while use of the 16,000-kilometer-long National Cycle Network rose by 15 percent between 2004 and 2005.[13]

In Australia, the state of Victoria committed in 2006 to triple its funding of bikeways over the next decade.[14] France created the position of "Monsieur Velo"—a Cycling Czar—whose chief responsibility is to increase biking rates, likely through promoting measures to increase cycling safety, provide more bicycle parking, and offer greater bike access to public transportation.[15] And in San Francisco, the city has set a goal of raising commuters' bicycle use from 2 to 10 percent of trips by 2010.[16]

Achieving such a transformation in U.S. cities is a tall order. But a 2006 study found that the higher cycling rate in Canada—three times above that in the United States—could be traced to infrastructure differences rather than to weather or cultural differences.[17] It found, for example, that cycling rates in the Yukon are twice as high as in Southern California and three times as high as in Florida.[18] The authors concluded that cycling in the United States could be increased through changes in transport and land use policies.[19] U.S. transportation legislation has increased funding for bicycling infrastructure from $150 million in 1992–97 to $900 million for 2005–09.[20] But cycling's share of transportation funding remains minuscule.[21] (See Figure 3.)

Meanwhile, entrepreneurs continue to play a role in promoting cycling. A subscription bike rental service opened in Boston in 2006 that offers a year's use of a 15-speed mountain bike for just $19.99.[22] The bulk of the revenue for the service comes from advertising mounted on the bike frames, which costs companies about $100 a month for ads on four bikes.[23]

Figure 1. World Bicycle Production, 1950–2004

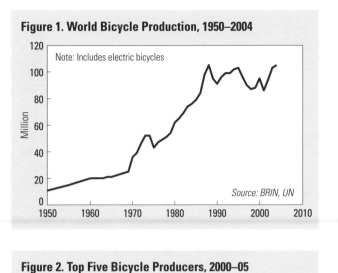

Note: Includes electric bicycles

Source: BRIN, UN

Figure 2. Top Five Bicycle Producers, 2000–05

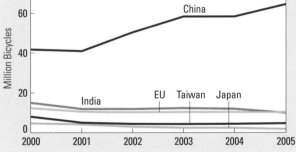

Source: BRIN

China

India   EU   Taiwan   Japan

Figure 3. Share of U.S. Transportation Funding by Mode, 2005

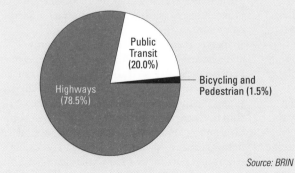

Public Transit (20.0%)

Bicycling and Pedestrian (1.5%)

Highways (78.5%)

Source: BRIN

World Bicycle Production, 1950–2004

Year	Production
	(million)
1950	11
1955	15
1960	20
1965	21
1970	36
1971	39
1972	46
1973	52
1974	52
1975	43
1976	47
1977	49
1978	51
1979	54
1980	62
1981	65
1982	69
1983	74
1984	76
1985	79
1986	84
1987	98
1988	105
1989	95
1990	91
1991	96
1992	99
1993	99
1994	102
1995	103
1996	96
1997	90
1998	87
1999	88
2000	95
2001	86
2002	94
2003	103
2004	105

Source: Bicycle Retailer and Industry News and United Nations.

Air Travel Reaches New Heights

Zoë Chafe

In 2005, the number of passengers traveling on scheduled airlines in a single year passed 2 billion for the first time, according to provisional estimates from the International Civil Aviation Organization (ICAO).[1] (See Figure 1.) And travelers flew an unprecedented distance— more than 3.7 trillion passenger-kilometers.[2] (See Figure 2.) This is equivalent to 4.8 million people flying to the moon and back in one year.

ICAO numbers are an underestimate of total plane travel, as they do not include military aviation or the private and business jet industry that is so popular with politicians, business people, and celebrities.[3] Some 4,000 new private and business planes were built in 2006, at a record cost of $18.8 billion, up 21 percent from 2005.[4] (See Figure 3.) Private planes emit up to four tons of carbon dioxide per hour and carry few people, so the pollution-per-passenger ratio is much greater than on commercial flights.[5]

Due to war, terrorism, disease outbreaks, and rising fuel costs, the global airline industry has not turned a profit since 2000, though the International Air Transport Association expects an industry-wide profit of $2.5 billion in 2007, a profit margin of 0.5 percent.[6] All regions except Africa are expected to be profitable.[7]

As the number of passengers flying each year grows, so does aviation-related pollution. One airplane crossing the Atlantic can use 60,000 liters of fuel—about as much as a driver uses in 50 years.[8] The Intergovernmental Panel on Climate Change reports that in 1992, the last year with data available, air transport contributed 2 percent of global carbon dioxide emissions but nearly all of the nitrogen oxide emissions found 8–15 kilometers above Earth.[9] Because the emissions from planes occur so high in the atmosphere, they contribute to global warming at two to four times the rate of emissions closer to Earth, such as those from cars.[10]

Currently, emissions produced on international flights are not explicitly regulated by the Kyoto Protocol, though those produced during domestic flights are included in country-specific targets.[11] The ICAO expects to issue guidelines for emissions trading related to international

LINKS pp. 42, 54

aviation in September 2007.[12]

Within the European Union (EU), by 2011 airlines will be accountable for emissions from all domestic air travel and flights between member countries under the EU Emissions Trading Scheme (ETS).[13] And in 2012, all flights arriving at or departing from an EU airport will be subject to ETS, with caps set at the average level of emissions between 2004 and 2006.[14]

One way to reduce the emissions caused by flying is to improve international air traffic management. Cutting flight times and making routes more efficient would avoid an estimated 73 million tons of carbon emissions each year.[15] Virgin Air is experimenting with electric tractors that tow planes from gates to the runway, saving up to 2,500 liters of fuel per flight.[16]

Aircraft design matters immensely: new airplanes are 70 percent more fuel-efficient than those designed 40 years ago and 20 percent more efficient than those built just 10 years ago.[17] High fuel prices provide a continued incentive to design more-efficient planes. Future planes may have longer, lighter wings with engines mounted at the tips, which would reduce drag and cut plane weight.[18]

Noise and material waste are also significant environmental concerns. The Natural Resources Defense Council reports that the U.S. airline industry throws away enough aluminum cans each year to build 58 Boeing 747 airplanes.[19] Paper is the largest category of waste generated by the industry.[20]

Some airports are taking their environmental records seriously. Japan's Centrair Airport, opened in 2005, was built on an artificial island to minimize noise and air pollution over nearby communities.[21] The island was carefully shaped to preserve existing ocean currents, and a hydrogen fuel cell bus ferries passengers between terminals.[22] In 2006 Boston's Logan Airport Terminal A became the first U.S. airport to receive the Green Building Council's Leadership in Energy and Environmental Design certification for, among other features, its natural lighting, energy-saving roof design, and use of paints and sealants with low volatile organic compounds.[23]

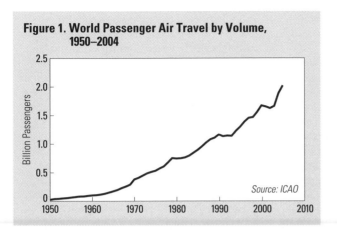

Figure 1. World Passenger Air Travel by Volume, 1950–2004

Source: ICAO

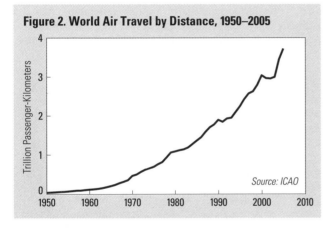

Figure 2. World Air Travel by Distance, 1950–2005

Source: ICAO

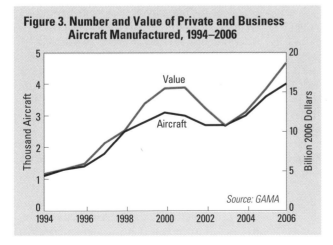

Figure 3. Number and Value of Private and Business Aircraft Manufactured, 1994–2006

Source: GAMA

World Air Travel by Distance and Passenger Volume, 1950–2005

Year	Passengers	Distance
	(million)	(billion passenger-kilometers)
1950	31	28
1955	68	61
1960	106	109
1965	177	198
1970	383	460
1975	534	697
1976	576	764
1977	610	818
1978	679	936
1979	754	1,060
1980	748	1,089
1981	752	1,119
1982	766	1,142
1983	798	1,190
1984	848	1,278
1985	899	1,367
1986	960	1,452
1987	1,028	1,589
1988	1,082	1,705
1989	1,109	1,774
1990	1,165	1,894
1991	1,135	1,845
1992	1,146	1,929
1993	1,142	1,949
1994	1,233	2,100
1995	1,304	2,248
1996	1,391	2,432
1997	1,457	2,573
1998	1,471	2,628
1999	1,562	2,798
2000	1,674	3,038
2001	1,655	2,966
2002	1,627	2,957
2003	1,665	2,998
2004	1,888	3,445
2005	2,022	3,720

Source: ICAO.

Cell Phones Widely Used, Internet Growth Slows   Zoë Chafe

The world now has more than 2 billion cell phone or mobile phone subscribers, according to the latest data from the International Communications Union.[1] (See Figure 1.) Some 410 million new cell phone subscribers signed up in 2005 alone, more than the combined population of the United States and Canada.[2] The number of subscribers climbed an average of 24 percent each year over the past five years.[3] At the same time, the number of fixed telephone lines (known now as land lines) has almost stagnated, reaching 1.3 billion in 2005, with an average growth rate of 5 percent over the past five years.[4]

The number of cell phone users per 100 residents varies drastically by country. In Israel and the United Kingdom, there are more cell phone subscriptions than there are people.[5] Japan, South Africa, and the United States have similar rates of cell phone subscribers—about 70 per 100 residents.[6] The two most populous countries in the world have much lower rates, with 30 subscribers per 100 residents in China and 8 per 100 in India.[7]

Cell phones are increasingly used in ways that have little to do with their original function. In Japan, people can now pay for food and train tickets with their cell phones.[8] They can also scan barcodes on fresh produce packaging, instantly retrieving information about where the food was grown and whether pesticides were used.[9]

Some creative uses are more crucial to basic survival. In Bangladesh, the Welltracker project helps villagers ensure the safety of their water supply by phone: after sending a series of messages to pinpoint their location, they receive information from a database about how deep they should dig their well in order to avoid arsenic contamination.[10] And WeatherBug, a U.S. company, has announced a service that sends severe weather alerts based on a cell phone user's location.[11] Weather sites are the second most popular category of Web site, after e-mail, visited by people who get on the Internet via their phones in the United States.[12]

There were an estimated 1.2 billion Internet users worldwide in 2006, up 13 percent over 2005.[13] While the Internet is widely available and relatively cheap to use in some places, the percent of the population that goes online varies greatly between countries. Iceland has the highest concentration of Web users, at 87.8 percent, followed by Sweden (75.5 percent) and Australia (70.4 percent).[14] (See Figure 2.) But in 97 countries, fewer than 10 people per 100 residents use the Internet; this includes 29 countries where the figure is below 1 in 100.[15]

The number of Internet host computers grew by 38 million to a total of 433 million computers in 2006, but this represented the slowest annual growth rate (9.7 percent) since surveys began in 1985.[16] (See Figure 3.)

Though computers have become an integral part of many lives, few people realize the toxic burden they carry. Nearly one kilogram of a typical laptop computer—about 23 percent of its weight—is composed of metals that can be harmful to humans in high concentrations, such as lead, cadmium, and copper.[17] One ton of discarded computers has more gold than is produced from 17 tons of gold ore.[18] While these metals may not be directly harmful to computer users, they have dire effects for the thousands of people worldwide who work as electronics recyclers, many without proper equipment or protection, to process the estimated 20–50 million tons of electronic waste generated each year.[19]

Internet users often take unfettered Web access for granted. But some national regulation threatens the integrity of the globe-crossing technology. In China, government censors intermittently shut down access to selected Web sites. The popular search engine Google, known as *Gu Ge* or "harvest song" in Chinese, has come under fire for bowing to Chinese government pressure and restricting search results for sensitive topics such as human rights and political reform.[20]

One founder of the World Wide Web, Sir Tim Berners-Lee, warned that a recent proposal could cause the Internet to enter "a dark period."[21] Large U.S. telecom businesses want to grant subscription-only access to parts of the Internet, with priority given to data transmitted by companies or institutions that pay higher rates.[22]

Figure 1. Cellular Telephone Subscribers Worldwide, 1985–2005

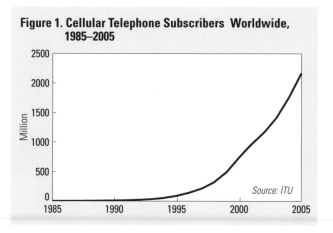

Source: ITU

Cellular Telephone Subscribers and Internet Host Computers Worldwide, 1985–2006

Year	Cellular Phone Subscribers	Internet Host Computers
	(million)	
1985	1	0.00
1986	1	0.01
1987	2	0.03
1988	4	0.08
1989	7	0.16
1990	11	0.38
1991	16	0.73
1992	23	1.31
1993	34	2.22
1994	56	5.85
1995	91	14.35
1996	145	21.82
1997	215	29.67
1998	319	43.23
1999	492	72.40
2000	740	109.57
2001	964	147.34
2002	1,160	171.64
2003	1,411	233.10
2004	1,759	317.65
2005	2,168	394.99
2006 (prel)	n.a.	433.19

Source: ITU, Internet Systems Consortium.

Figure 2. Top 10 Countries by Internet Users Per 100 People, 2005

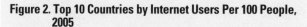
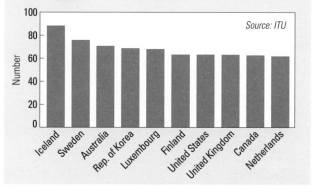

Source: ITU

Figure 3. Number of Internet Host Computers Worldwide, 1985–2006

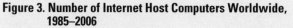
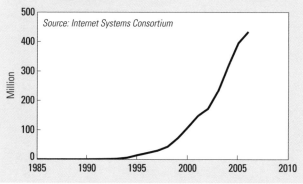

Source: Internet Systems Consortium

Conflict and Peace Trends

A young man of the White Army with his heavy machine gun, Akobo, Sudan.

© UN/IRIN

▶ Number of Violent Conflicts Steady

▶ Peacekeeping Expenditures Hit New Record

▶ Nuclear Weapons Treaty Eroding

For data and analysis on conflict and peace trends, including military expenditures and resource conflicts, go to www.worldwatch.org/vsonline.

Number of Violent Conflicts Steady

Michael Renner

During 2006, the number of wars and armed conflicts worldwide remained almost unchanged from previous years at 43, according to AKUF, a conflict research group at the University of Hamburg.[1] (See Figure 1.) There were 28 full-fledged wars and 15 armed conflicts (organized violence of lesser severity).[2] Asia was home to the most conflicts (16), followed by Africa (12) and the Middle East (11).[3]

Three conflicts were terminated, yet elsewhere an equal number of new conflicts erupted. Fighting ended in Indonesia (Aceh), India (Bodos), and Ethiopia (Gambela region).[4] But East Timor, the Central African Republic, and Brazil (where criminal violence in São Paulo escalated) were subjected to fresh outbreaks of violence.[5] And the scale of violence increased sufficiently during 2006 in Sri Lanka and in Pakistan's Baluchistan province to merit reclassification from armed conflicts to full wars.[6]

Available information about many conflicts is typically incomplete and often even contradictory. In addition, varying definitions and methodologies—sets of criteria that need to be met in order for episodes of violence to be counted as a conflict—are used by different researchers. It is thus not surprising that such efforts can lead to diverging assessments.

The Uppsala/PRIO Armed Conflict Dataset, one of the most widely used, indicates conflict trends that follow a roughly similar trajectory as the AKUF data. However, the aggregate numbers of active conflicts for any given year are quite different.[7] (See Figure 2.)

The Heidelberg Institute for International Conflict Research (HIIK) in Germany finds that while the number of "high intensity" conflicts has indeed declined in recent years, "medium intensity" conflicts (characterized by sporadic rather than continuous violence) have risen steeply, as have "low intensity" (nonviolent) conflicts.[8] (See Figure 3.) Some 58 percent of all conflicts in 2006 involved nonviolence.[9]

Although reliable data are scarce, the Democratic Republic of the Congo, Sudan's Darfur region, and Iraq rank among the deadliest conflict areas.[10] For example, a rigorous survey of mortality in Iraq after the U.S. invasion, based on interviews with close to 2,000 households across the country, led researchers to extrapolate that some 600,000 Iraqis may have died violent deaths between March 2003 and July 2006.[11] This finding proved controversial, particularly since other estimates are far lower.[12] Yet other tallies are based on records such as media reports or mortuary and hospital death data that for a variety of reasons fail to capture a significant portion of total deaths.

Wars not only kill, they drive many people out of their homes—forcing them to flee to other countries or to seek safety elsewhere within their own country. In an encouraging trend, the U.N. High Commissioner for Refuges reports that the number of refugees at the beginning of 2006 was at its lowest level since 1980—8.4 million.[13] During 2005, some 1.1 million mostly Afghan refugees returned home voluntarily, while only 136,000 people newly became refugees—the smallest number in 29 years.[14]

However, the number of so-called internally displaced persons (IDPs)—those who do not cross an international border and often lack even the most rudimentary protections—has not declined. The World Refugee Survey put the number of IDPs worldwide at 20.1–21.3 million as of December 2005.[15] The Internal Displacement Monitoring Centre offers an estimate of 23.7 million.[16] Among the most worrisome situations is the one in Iraq, with some 1.7 million IDPs and 2 million refugees, or 13 percent of the population.[17]

A broad range of conflict resolution efforts were undertaken in 2006. In at least 31 of the 278 active conflicts tallied by HIIK, talks or negotiations were held.[18] Six peace treaties and seven ceasefire agreements were signed.[19] The United Nations and other entities carried out more than 60 peacekeeping and monitoring missions.[20] A variety of sanctions are in force in order to contain or end conflicts. The United Nations maintained sanctions against eight states during 2006 (Côte d'Ivoire, Democratic Republic of the Congo, Liberia, Rwanda, Sierra Leone, Somalia, Sudan, and North Korea), as well as al-Qaeda and the Taliban in Afghanistan.[21]

Figure 1. Wars and Armed Conflicts, 1950–2006

Source: AKUF

Figure 2. Armed Conflicts, 1950–2005

Source: Gleditsch et al.

Figure 3. Conflicts by Intensity Level, 1950–2006

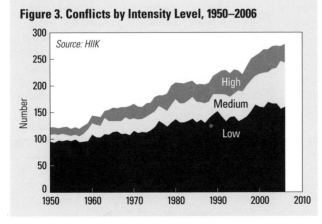

Source: HIIK

Wars and Armed Conflicts, 1950–2006

Year	Wars	Wars and Armed Conflicts
		(number)
1950	13	
1955	15	
1960	12	
1965	28	
1970	31	
1975	36	
1976	34	
1977	36	
1978	37	
1979	38	
1980	37	
1981	38	
1982	40	
1983	40	
1984	41	
1985	41	
1986	43	
1987	44	
1988	45	
1989	43	
1990	50	
1991	54	
1992	55	
1993	48	62
1994	44	59
1995	34	48
1996	30	49
1997	29	47
1998	33	49
1999	35	48
2000	35	47
2001	31	48
2002	29	47
2003	27	43
2004	27	42
2005	28	42
2006 (prel)	28	43

Source: AKUF.

Peacekeeping Expenditures Hit New Record   *Michael Renner*

The approved budget for United Nations peace-keeping operations from July 2006 to June 2007 reached an unprecedented $5.28 billion.[1] (See Figure 1.) The United Nations now also deploys more soldiers, military observers, and police than ever before—peaking at 82,120 in November 2006.[2] (See Figure 2.) When international and local civilian staff and volunteers are included, along with the mostly civilian staff of a number of smaller "political and peace-building" missions, U.N. peacekeeping staff crosses the 100,000 threshold for the first time.[3]

With a newly beefed-up Lebanon mission growing toward its authorized strength of 15,000 and a Darfur mission in the cards, the uniformed peacekeeping ranks may soon rise past 115,000, with annual costs of $7 billion.[4] Yet the U.N. peacekeeping department, starved of adequate resources by its member states, worries that the strains already apparent may turn into fatal overstretch.[5] Over the department's objections, the Security Council renewed its call for an 11,000-strong force to protect refugees along the Chad-Sudan border.[6]

Although they are at historic highs, U.N. peacekeeping budgets continue to be dwarfed by military expenditures. According to the Stockholm International Peace Research Institute, military budgets stood at $1,118 billion in 2005—212 times as much as spent on U.N. peacekeeping.[7]

LINKS   p. 76

A total of 114 countries contributed personnel to U.N. missions in 2006.[8] But the nations of the Indian subcontinent—Pakistan, Bangladesh, India, Nepal, and Sri Lanka—accounted for 44 percent.[9] Ethiopia, Ghana, Nigeria, South Africa, Senegal, and Morocco provided another 18 percent.[10] By contrast, the powerful permanent members of the Security Council—China, France, Russia, the United Kingdom, and the United States—contributed a mere 4.6 percent of all personnel.[11]

Six missions account for about 80 percent of current peacekeeping budgets.[12] Missions in the Democratic Republic of the Congo and in Sudan each cost about $1.1 billion in 2006–07.[13] The next four largest deployments—in Liberia, Haiti, Côte d'Ivoire, and Lebanon—together

cost about $2.1 billion.[14] These top six missions also account for 86 percent of all personnel.[15]

In East Timor, peacekeepers returned in August 2006. An earlier mission that had facilitated the country's independence from Indonesia was scaled down and withdrawn in May 2005—prematurely, it turned out, when violence broke out once more in early 2006.[16] There is some worry that a mission ended in Burundi in December 2006 may eventually suffer a similar fate—the government asked the United Nations to leave ahead of schedule.[17]

Peacekeeping efforts remain under a financial cloud, as member states continue to pay their dues late or not in full. As of October 2006, peacekeeping arrears stood at $2.5 billion, the third-highest year-end outstanding sum ever.[18] Two nations were responsible for more than half the debt: the United States with $800 million and Japan with $563 million in arrears.[19] France owed $174 million and China $110 million, while the next 11 largest contributors together owed another $443 million.[20]

Non-U.N. missions can also be found in all regions of the world, sometimes working in conjunction with the U.N. "Blue Helmets." During 2006, 47 missions were maintained by regional organizations such as the European Union and the Organization for Security and Co-operation in Europe or by ad hoc coalitions.[21] Altogether, they involved an estimated 50,000 soldiers—down from a peak of 115,000 in 1999, when North Atlantic Treaty Organization deployments in the Balkans reached 70,000.[22] (See Figure 3.) Thus together with the U.N. missions, more than 150,000 people were deployed for peacekeeping in 2006.[23]

Peacekeeping mandates range from monitoring and policing to muscle-bound "peace enforcement." Most U.N. operations nowadays are under Chapter 7 of the U.N. Charter (that is, they are authorized to use force).[24] This is an outgrowth of earlier missions that were seen as toothless. But there is always a danger that peacekeepers become just another warring party. And some missions by "coalitions of the willing" may be guided more by self-interest than an earnest desire for peace.

Figure 1. U.N. Peacekeeping Expenditures, 1950–2006

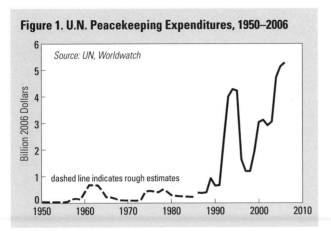

Source: UN, Worldwatch

dashed line indicates rough estimates

Figure 2. U.N. Peacekeeping Personnel, 1950–2006

Source: UN, Global Policy Forum, Stimson Center

Figure 3. Non-U.N. Peacekeeping Personnel, 1976–2006

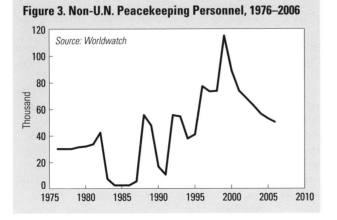

Source: Worldwatch

U.N. Peacekeeping Expenditures, 1986–2006

Year	Expenditure
	(billion 2006 dollars)
1986	0.378
1986	0.393
1987	0.380
1988	0.407
1989	0.935
1990	0.658
1991	0.672
1992	2.368
1993	4.006
1994	4.285
1995	4.227
1996	1.640
1997	1.201
1998	1.200
1999	1.948
2000	3.044
2001	3.131
2002	2.926
2003	3.068
2004	4.728
2005	5.137
2006 (prel)	5.280

Note: 1986–94 is calendar year; 1995–2006 is July to June of following year.

Source: U.N. Department of Public Information and Worldwatch Institute database.

Nuclear Weapons Treaty Eroding

Michael Renner

The number of nuclear warheads held in 2006 by the world's five full-fledged nuclear powers—the United States, Russia, the United Kingdom, France, and China—was estimated at about 27,000.[1] (See Figure 1.) Combined, these contain about 5,000 megatons of explosive material—a massive overkill capacity that could destroy human civilization.[2] Thankfully, however, the arsenal is down from a peak of about 70,000 in 1986 and is at its lowest level since 1961.[3]

Nuclear warheads can be delivered to distances near and far by almost 2,200 missiles and bombers, most of which carry multiple warheads.[4] Almost 20 years after the end of the cold war, about 2,500 of the world's nuclear warheads remain on hair-trigger alert, meaning they can be launched within minutes.[5]

The United States and Russia control about 97 percent of global nuclear arsenals.[6] (See Figure 2.) Of roughly 10,000 U.S. warheads, 5,735 are thought to be operational (with the remainder to be dismantled over many years).[7] Washington also maintains 5,000 plutonium cores, which can be turned into warheads, as a strategic reserve.[8] Of Russia's 16,000 warheads, 5,830 are estimated to be deployed.[9] France, China, and the United Kingdom are believed to have a combined force of roughly 750 warheads.[10] (See Figure 3.)

Since 1945, more than 128,000 warheads have been built: more than 70,000 by the United States; 55,000 by the Soviet Union or Russia; 1,200 by the United Kingdom; more than 1,260 by France; and some 600 by China.[11]

The Nuclear Non-Proliferation Treaty (NPT) bars additional countries from acquiring nuclear weapons and commits states with such arms to move toward complete nuclear disarmament.[12] But the norms enshrined in the NPT are increasingly being ignored.

On one side, the recognized nuclear powers refuse to live up to their commitments and are instead planning to modernize their arsenals. The United States is developing new warheads as well as new missiles and submarines to carry them and is gearing up to restart plutonium production.[13] Refurbishing the nuclear weapons manufacturing complex may cost more than

$100 billion.[14] Russia is introducing a new intercontinental ballistic missile, a new class of strategic submarines, and a new cruise missile.[15] China will soon deploy new long-range missiles.[16] France is developing nuclear-powered submarines armed with a new type of ballistic missile.[17] And the United Kingdom is planning to acquire a new generation of nuclear missile–carrying submarines at a projected cost of $40 billion.[18] Prime Minister Blair justified this project, which will take 17 years to complete, as a needed deterrent against North Korea and Iran.[19]

On the other side, India and Pakistan have acquired nuclear arms but remain outside the NPT. They are estimated to have built about 110 warheads between them and have sufficient fissile material for perhaps another 85–110.[20] Israel will not officially confirm that it possesses nuclear weapons, but experts estimate the country has 60–85 warheads and fissile material stocks that are equivalent to 115–190 warheads.[21]

North Korea announced in early 2005 that it possessed nuclear weapons and said in October 2006 that it had carried out a nuclear test.[22] Meanwhile, questions abound as to whether Iran's nuclear program is of a purely civilian nature or is intended to produce weapons.[23] When Iran rejected a call by the U.N. Security Council in July 2006 to halt its uranium enrichment program, the Council imposed limited sanctions in December.[24]

A distinct danger of escalating tensions remains. A variety of observers and analysts are afraid that the United States and Israel, acting alone or jointly, might conduct air strikes against Iran's nuclear facilities.[25]

Even as they warn other nations to renounce possession of nuclear arms, the five recognized nuclear powers continue to insist that their arsenals are indispensable to their security. This is an invitation to other governments to break out of the NPT regime. The integrity of the NPT received another blow when the United States signed a treaty on civilian nuclear cooperation with India that effectively allows that country to step up its nuclear arms manufacturing, even though the government still rejects the NPT.[26]

Figure 1. Global Nuclear Warheads, 1945–2006

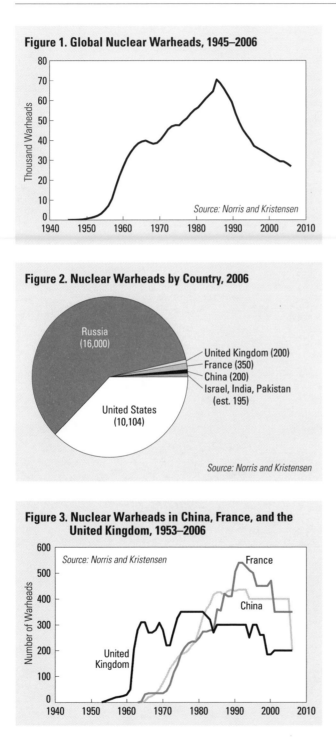

Source: Norris and Kristensen

Figure 2. Nuclear Warheads by Country, 2006

Russia (16,000)

United Kingdom (200)
France (350)
China (200)
Israel, India, Pakistan (est. 195)

United States (10,104)

Source: Norris and Kristensen

Figure 3. Nuclear Warheads in China, France, and the United Kingdom, 1953–2006

Source: Norris and Kristensen

France

China

United Kingdom

Global Nuclear Warheads, 1945–2006

Year	Warheads
1945	6
1950	374
1955	3,267
1960	22,069
1965	38,458
1970	38,696
1975	47,604
1976	47,536
1977	49,544
1978	51,024
1979	53,360
1980	55,246
1981	56,467
1982	58,629
1983	60,882
1984	62,753
1985	64,519
1986	70,481
1987	68,479
1988	65,726
1989	62,525
1990	59,239
1991	53,562
1992	49,022
1993	45,336
1994	42,715
1995	40,344
1996	37,159
1997	36,060
1998	34,981
1999	33,859
2000	32,632
2001	31,477
2002	30,425
2003	29,371
2004	29,308
2005	28,245
2006 (prel)	26,854

Source: Norris and Kristensen, "Nuclear Notebook."

Part Two

SPECIAL FEATURES

Food and Agriculture Features

A woman works in her irrigated urban garden in Addis Ababa, Ethiopia.

▶ Agribusinesses Consolidate Power

▶ Egg Production Doubles Since 1990

▶ Avian Flu Spreads

For data and analysis on these food and agricultural topics and on others such as illegal drug harvesting and farm subsidies, go to www.worldwatch.org/vsonline.

Agribusinesses Consolidate Power E. Starmer and M. D. Anderson

At all stages of the food system—from seeds and other inputs to food processing and retail food sales—market power is concentrating in an ever smaller number of corporate firms. This trend is transforming how the world produces food, squeezing millions of farmers between a small group of input suppliers and an equally concentrated group of commodity purchasers, and in turn influencing the food choices available to consumers.

Concentration begins at the input stage in agriculture. Three companies control about half of the global agrochemical market: Bayer, Syngenta, and BASF.[1] Use of genetically modified (GM) seeds has risen dramatically since these were first commercialized in the mid-1990s— now 45 percent of the corn and 85 percent of the soybeans grown in the United States are GM.[2] By branching out into plant biotechnology, huge chemical and pharmaceutical companies such as Monsanto have gained control over critical agricultural inputs that reach into food systems around the world. In 2004, land planted with Monsanto seeds accounted for 88 percent of the total area in GM crops worldwide.[3] Once a global commons, genetic resources are now subject to Intellectual Property Rights protections. Developing countries are forced to deal with large transnational companies to get access to improved seed varieties and plant breeding technologies.[4]

pp. 20, 22, 24, 54

Other input markets are similarly concentrated. In the United States, Mosaic—a company created out of a merger between Cargill and IMC Global—controls 50–60 percent of the synthetic fertilizer market, while four firms control over 80 percent of the market for farm equipment.[5] Four companies control 60 percent of terminal grain facilities, and Cargill, Archer Daniels Midland, and Zen Noh control 81 percent of U.S. corn exports and 65 percent of soybean exports.[6] Cargill has the largest global terminal capacity, handling significant grain exports in Canada, the United States, Brazil, and Argentina.[7] It owns and operates a worldwide transportation network of ships, trucks, barges, railcars, and grain elevators for storage.

Cargill is also among the top three beef producers in the United States and plays an important role in poultry production.[8]

Genetic stock, a crucial input into animal production, is another area where concentration has rapidly taken hold.[9] Control over the development and dissemination of livestock genetics is shifting from farmers and ranchers to specialized genetics companies. They hold exclusive contracts with the largest livestock producers and play a key role in determining which livestock breeds will dominate the market.[10] Today, virtually all white eggs sold on the U.S. market come from a single breed of layer, the white leghorn.[11] A depleted genetic pool will weaken the global food system's ability to respond to disease, to changes in climate or available inputs, and to shifts in consumer preference.[12]

A growing share of farmers and ranchers in the United States, Europe, and some developing countries work under contract for companies that also control food processing and distribution. These firms may mandate the use of a certain technology to maximize yield or animal weight gain. If producers stray from the prescribed methods, they may find their contracts terminated.[13] Virtually all U.S. poultry is produced under contract, as are close to 60 percent of hogs, cotton, rice, fruit, and dairy.[14] Contracts tend to shift risk from the company to the producer, and producers are often forced by necessity into contracts that pay little or are excluded altogether from markets if they do not contract.[15]

Whether producing independently or under contract, farmers have few choices when it comes to selling their product to a packer or processor. In Brazil, 68.5 percent of the soybean oil refineries are controlled by just three companies.[16] In the United States, 81 percent of beef packing plants are run by four firms.[17] (See Table 1.) Concentration in livestock and dairy markets is likely to continue in developing countries as well, as rising incomes and shifting dietary preferences boost meat consumption.[18]

Globally, transnational supermarkets dominate the retail sector for food. In 2003, the top

Table 1. Share of U.S. Processing and Packing Markets Held by Top Four Firms, 2005

Market	Share
	(percent)
Beef packing	81
Pork packing	59
Broiler production and processing	50
Turkey production and processing	45
Flour milling	61
Soybean crushing	80

Source: Hendrickson and Heffernan.

30 retailers held 19 percent of the market in Asia and Oceania, 29 percent of the market in Latin America, and 69 percent of the market in Europe.[19] Globalized supply chains give supermarkets the ability to get products from wherever they are cheapest, and the large firms exert pressure on suppliers to accept lower prices. Suppliers in turn demand that farmers accept lower prices. Squeezed between low returns and high-priced farm inputs, farmers around the world have experienced declines in net farm income. In the United States, farmers' share of the retail food dollar fell from a high of 40 percent in 1973 to below 20 percent in 2000.[20] In Canada, the National Farmers Union reported that farmers' net income, adjusted for inflation, was lower over the last decade than at any time since the 1930s.[21]

Some analysts argue that large supermarkets like Wal-Mart's Supercenters have helped consumers by using market power to drive down prices.[22] But a growing body of economic research suggests that, over time, concentration tends to lead to higher prices.[23] Because of the power they exert over the market, giant retailers have no incentive to pass on savings to consumers, even as they squeeze producers and suppliers by offering lower and lower prices for their products.[24]

In a striking example of the power of large processors and retailers, U.S. hog prices fell to Depression-era lows in real terms in 1998, sending many family hog producers into bankruptcy.[25] Meanwhile, the average price of pork in the grocery store dipped by less than 2 percent.[26] This wide farm-retail price spread helped the giant meatpacking company IBP bring in record profits and facilitated market dominance by industrialized hog operations.[27]

Around the world, individuals, communities, and civil society organizations are working to counteract the negative impacts of concentration in the food system. In the United States, they are trying to strengthen existing laws, such as the Packers and Stockyards Act, that have been weakened by lax enforcement, underfunding, or legal loopholes.[28]

Campaigns against abusers of market power are taking shape. In Europe, a major campaign has been launched against the largest supermarket, Tesco. It demands fair treatment of U.K. farmers and those abroad; protection of workers' rights; an independent watchdog agency to protect consumers, farmers, and workers against exploitation; a moratorium on mergers with other supermarkets; and stronger planning policies to protect local shops.[29] Organizations are using class action lawsuits and penalties against retail giant Wal-Mart for discrimination against women, forced overtime without pay, abuse of Family Leave laws, and other labor problems.[30] International networks such as the Agribusiness Accountability Initiative are helping campaigners to connect across national boundaries.

For farmers, the most effective strategy is strength in numbers: forming cooperatives so that they can supply enough reliable quantity and quality of crops or livestock products to negotiate with supermarkets. At the same time, public education campaigns worldwide are raising awareness about direct marketing options for farmers and consumers, including farm stands, farmers' markets, and Internet sales. But farmers need government support to keep agribusinesses in check and to meet the quality standards that these large companies impose.

Egg Production Doubles Since 1990

Katie Carrus

Global egg production doubled between 1990 and 2005.[1] By then, some 64 million tons of eggs were produced worldwide (less than 1 percent more than in 2004).[2] (See Figure 1.) Today there are approximately 4.93 billion egg-laying hens in the world, each capable of producing up to 300 eggs per year.[3] By 2015, world egg production is expected to reach 72 million tons, according to the U.N. Food and Agriculture Organization (FAO).[4]

While egg production has increased in the United States, Japan, India, and Mexico over the past four decades, most of the growth has been due to a 10-fold increase in eggs in developing countries in response to rising incomes and growing populations.[5] Between 1990 and 2005, China accounted for 64 percent of the growth in world egg production.[6] By 2005 this one country produced nearly 44 percent of the world's eggs—28.7 million tons—more than five times as many as the next largest producer.[7] (See Figure 2.) And this trend is expected to continue, with output there predicted to rise by 23 percent by 2015.[8]

By 2000, developing countries in Asia were producing twice as many eggs as all industrial countries.[9] Output in the United States grew 13 percent between 1995 and 2000, compared with 34 percent in China during the same period.[10] And in some countries, such as the United Kingdom, Japan, Hungary, and Denmark, fewer eggs were produced in 2000 than in 1998.[11] The growth rate throughout the industrial world between 1961 and 2000 was quite low: 1.6 percent.[12] Over the next 15 years, egg production in the industrial world is expected to increase from 18 million to 20 million tons, due in part to food saturation and overconsumption.[13]

People in industrial countries eat about twice as many eggs as people in developing countries—approximately 226 eggs per person per year.[14] Yet only 30 countries are seeing any growth in per capita egg consumption.[15] Among these nations are China, Libya, Mexico, Colombia, Turkey, and India.[16] Elsewhere, egg consumption is either stable or falling.[17] FAO predicts that most future growth in egg con-

LINKS pp. 24, 90

sumption will occur in the developing world in places like China, where income and population patterns are still shifting.[18]

Most egg production in China has transitioned from traditional, scattered, backyard farms to large-scale integrated operations.[19] While small farmers once produced most of the eggs for markets for local consumers, large-scale, vertically integrated factory farming has become the norm. Producers now typically confine egg-laying hens in small wire "battery" cages stacked in rows in sheds that are the length of a football field.[20] Indeed, nearly 60 percent of China's egg production in 2005 was done on farms with more than 500 layers.[21] Taiwan alone produced about 390,000 tons of eggs in 2005 on 1,400 facilities housing on average 40,000 birds each.[22]

Market concentration and industrial, intensive production methods like these have found favor among Chinese egg industry leaders.[23] "Intensification promises to be the right track for China to follow to develop its poultry industry," noted Hongge Wang, senior economic expert for animal husbandry at the China National Animal Husbandry and Veterinary Service in Beijing.[24] The Chinese government has already developed policies to encourage this, such as subsidies for large-scale farms.[25]

These policies have troubling implications for the environment, human health, and animal welfare. The Chinese State Environmental Protection Administration reports that industrial animal farms have become a major source of pollution, with raw manure being dumped into rivers that are a source of drinking water.[26] By 2002, Taihu Lake—a critical part of the Yangtze River delta—had become severely polluted with nitrogen and phosphorus from the untreated waste of industrial poultry farms.[27]

Avian influenza has ravaged much of the Asian poultry industry since 2003, with egg layer flocks often being more affected than broiler (meat chicken) flocks.[28] During the first four months of 2006, a commercial layer chick in China on average cost 24¢, a 12.6-percent decrease from the same period in the preceding year due to bird flu–related market disruptions.[29] In Thailand, efforts to stop the spread of avian flu led to the destruction of almost half of the country's 30 million egg-laying hens between November 2003 and February 2004.[30]

The industrial-style, intensive confinement of egg-laying hens in Asia has been strongly implicated in the epidemic's spread. The International Food Policy Research Institute notes that "the critical issue is the keeping [of] more and more animals in smaller and smaller spaces."[31] According to FAO, "once high-density industrial poultry areas become affected, infection can be explosively spread within the units, and the very high quantities of virus produced may be easily carried to other units, to humans, and into the environment."[32]

In addition, the intensive production methods that have enabled the dramatic surge in global egg production fall far short of any reasonable standard for animal welfare. Most hens on factory-style farms around the world live their entire lives in battery cages that frustrate most of their basic natural behaviors, including spreading their wings, walking freely, and nesting.[33] Due to growing consumer and governmental awareness of inherent animal welfare problems with the cage system, this production method is being phased out in the European Union.[34] And a growing consumer movement in the United States is steadily encouraging better standards for the country's 300 million egg-laying hens.[35]

Similar efforts are under way in Asia, but less regulated markets there have caused the factory farm egg industry to grow.[36] As public support for intensive confinement practices begins to dwindle in the West, large-scale egg producers are looking to Asia, where they can conduct business with little interference from

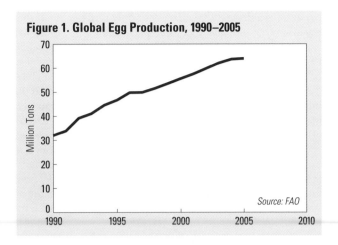

Figure 1. Global Egg Production, 1990–2005

Source: FAO

Figure 2. Top Egg Producing Countries, 1990–2005

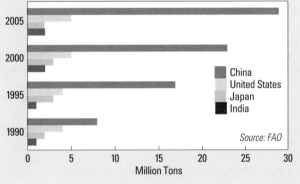

Source: FAO

individuals and groups concerned about animal welfare and environmental impacts.[37]

Avian Flu Spreads

Danielle Nierenberg

Avian flu is a disease that affects not only birds but also other animals, including pigs, cats, and humans. Since the latest major outbreak began in late 2003, at least 285 people have contracted the virus and 170 of them have died.[1] It is highly likely that other cases of infection and death have gone unreported, making it impossible to know the true scope of the disease.

Hundreds of millions of chickens, ducks, and other birds in areas where outbreaks occurred have been killed in an effort to control avian flu.[2] The virus, labeled H5N1, first jumped the species barrier in 1997 and spread to humans in Hong Kong, killing 6 of the 18 people infected.[3] The latest outbreak has now spread to more than 50 countries—nearly 40 of which were affected just in 2006—including China, India, Indonesia, Nigeria, and the United Kingdom.[4] (See Table 1.) Most of the human and avian deaths have occurred in Asia.[5]

In places with high concentrations of domestic pigs and chickens, pigs may serve as a "mixing vessel" for the virus because of their genetic similarity to humans.[6] In China, for example, where half of the world's pork is produced and consumed, pigs and chickens often live close to one another and to people in backyards or on factory farms.[7] The avian influenza virus could combine with pig influenza to create an entirely different strain of the disease, which could then be spread from coughing pigs to pig handlers and processors.[8]

LINKS pp. 24, 86, 88

Avian flu can also be spread directly from birds to humans through direct handling of chickens and the slaughtering and processing of meat—although not, experts say, from eating cooked meat from infected birds.[9] According to the World Health Organization (WHO), the current outbreak of H5N1 has been the most deadly of all the influenza viruses that have spread from birds to humans, killing more than half of the people infected—most of them previously healthy children and young adults.[10] Scientists are even more concerned that H5N1 will mutate into a virus that can easily spread from person-to-person, sparking a pandemic.[11]

If the disease does become a pandemic, loss of human life could be staggering. Thanks to globalized air travel and trade, a highly pathogenic flu virus could spread to every corner of the globe in just a matter of months. A December 2006 study in *The Lancet* estimated that as many as 62 million people could die in such a flu pandemic.[12] WHO gives a much more conservative number—estimating anywhere from 2 million to 7.4 million deaths from a pandemic.[13] But millions more would be forced to stay home from work, causing widespread economic disruption.[14] The World Bank estimates that current losses to the global poultry sector from avian flu are in excess of $10 billion.[15] If the virus becomes a pandemic, it could cost from $800 billion a year to $2 trillion overall.[16]

Developing nations will likely experience the greatest numbers of deaths because they lack access to vaccines or antivirals.[17] Governments are being encouraged to stock medications, such as the antiviral Tamiflu, to help combat an outbreak should it occur. But pharmaceutical companies lack the capacity—and often the financial incentive—to produce large numbers of these drugs quickly.[18] Currently, the World Bank estimates that $1.2–1.5 billion is needed over the next two to three years to address the financing gap for programs on avian and human influenza.[19]

In an attempt to deal with avian flu on the ground, at least 15 nations have restricted or even banned free-range and backyard production of birds, although this endangers the livelihoods of countless small farmers and jeopardizes the availability of affordable food for poor consumers.[20] Yet although migratory birds and small backyard farmers have been blamed for the spread of the disease, recent studies in Asia and Africa indicate that the real culprits may be factory farming and the globalized poultry trade and transport of livestock.[21]

Rising demand has helped drive livestock production from rural mixed farming systems, where farmers raise a few different species of animals on grass, to intensive periurban and urban production of pigs and chickens.[22] These confined animal feeding operations (CAFOs), or factory farms, create the perfect environment for the rapid spread of disease between animals and to people.[23] Because of unregulated zoning and subsidies that encourage livestock production, chicken and pig CAFOs are moving closer to major urban areas in China, Bangladesh, India, and many countries in Africa.[24] In Asia alone more than 6 billion birds are raised for food, many of them near the region's rapidly growing cities.[25]

In Laos, according to the Barcelona-based agriculture organization GRAIN, 42 of the 45 outbreaks of avian flu in spring 2004 occurred on factory farms and 38 of them were in the capital, Vientiane.[26] The few small farms where outbreaks occurred were located close to commercial operations. In Nigeria, the first cases of avian flu were found in one of the nation's industrial broiler operations.[27] The virus spread from that 46,000-bird farm to 30 other factory farms in the country—and then quickly to neighboring farms, forcing farmers to kill their chickens.[28] In India, the world's fifth largest producer of eggs and a leading producer of broiler chickens, the first outbreak occurred on a large factory farm.[29]

Avian flu has been found on backyard farms for centuries but has never been found to evolve there to the highly pathogenic form, such as the H5N1 virus.[30] Backyard poultry tend to be more genetically diverse and there are far fewer birds than in factory farms. These chickens receive more exposure to sunlight and have better ventilation, hygiene, and less stress than factory-farmed chickens—making them more resistant to disease.[31] Even genetically diverse native chickens cannot remain immune to the virus for long, however; it circulates from factory farms to backyard flocks and then back to factory farms, becoming more virulent.[32] Although having birds concentrated together in large factory farms may make it easier to monitor chickens and eradicate flocks, free-range birds are less likely to encourage an outbreak in the first place.[33]

And the avian flu virus continues to change. In 2004 some studies showed that the disease was becoming more lethal with every outbreak,

Table 1. Human Cases of Bird Flu and Deaths, by Country, March 2007

Country	Cases	Deaths
Azerbaijan	8	5
Cambodia	6	6
China	24	15
Djibouti	1	0
Egypt	32	13
Indonesia	81	63
Iraq	3	2
Thailand	25	17
Laos	2	2
Nigeria	1	1
Turkey	12	4
Viet Nam	93	42
Total	288	170

Source: WHO.

but a 2005 study maintained that some strains of the virus could be becoming "less virulent and more infectious," meaning that while it is not as lethal it could affect many more people.[34]

Despite bans on raising chickens outdoors, farmers will continue to do this in order to survive. Experts suggest that the Food and Agriculture Organization, WHO, and other international agencies should focus most of their avian flu prevention efforts on big poultry producers and on stopping disease outbreaks before they occur.[35] The industrial food system not only threatens the livelihoods of small farmers, it potentially puts the world at risk for a pandemic. Reversing this trend, according to GRAIN and other public interest groups, will mean standing up for farmers and demanding food production that is safe for animals and humans alike.[36]

Environment Features

Laysan albatrosses nest amid marine debris, Midway Atoll.

Steven Siegel, Marine Photobank

▶ Climate Change Affects Terrestrial Biodiversity

▶ Threats to Species Accelerate

▶ Invasive Species Drive Biodiversity Loss

▶ Ocean Pollution Worsens and Spreads

▶ Bottled Water Consumption Jumps

▶ Sustainable Communities Become More Popular

For data and analysis on these environmental topics and on others such as ecosystem stress, mangrove forests, and mercury pollution, go to www.worldwatch.org/vsonline.

Climate Change Affects Terrestrial Biodiversity   *Kevin P. Eckerle*

In its 2007 report, the Intergovernmental Panel on Climate Change (IPCC) left no doubt that global warming is occurring and that climate change is human-induced, concluding that "warming of the climate system is unequivocal" and stating with 90 percent confidence that the net effect of human activities on Earth since 1750 has been warming.[1] And it is increasingly clear that this warming climate is having significant impacts on the world's biodiversity.

In 2005 the Millennium Ecosystem Assessment (MA), which involved 1,360 scientists from 95 countries, concluded that climate change has affected biodiversity in all ecosystems over the last century, though the magnitude of the changes varied across ecosystem types.[2] (See Table 1.)

The most studied and best understood impact is phenological changes, which are alterations in the timing of periodic biological events, such as the onset of animal migration or plant blooming, in response to climatic conditions. These events are typically linked to climate and are predicted to occur increasingly early in response to Earth's steady warming; unfortunately, an increasing number of scientific studies present evidence consistent with this prediction. In plants, for instance, the flowering of cherry trees at the Royal Court in Kyoto, Japan, for which records have been maintained for at least 600 years, has advanced steadily since 1952.[3] And in the western United States, the flowering of lilacs and honeysuckles has advanced by 2 and 3.8 days per decade, respectively.[4] Increased warming has also extended the growing season of some plants, as in the eastern deciduous forest of the United States, where the growing season has lengthened most notably since 1966, and in the colder, most northerly zones at 42°–45° latitude.[5]

Among invertebrate animals, a study of 35 butterfly species in the United Kingdom found that the date of first appearance for 26 species has grown earlier by between 1.0 and 15.8 days per decade between 1976 and 1998, while for the remaining 9 species it either has not changed (2 species) or has gotten later by between 0.1 and 3.6 days per decade.[6] Likewise, between 1988 and 2002, the dates of first appearance for 17 Spanish butterfly species have advanced, as have the dates of first flight for 16 of 23 butterfly species in central California over 31 years.[7] In vertebrate animals, studies have documented that the males of four American frog species are initiating calling 10–13 days earlier than before and that migrant birds in the North Sea have been passing 0.5–2.8 days earlier per decade since 1960.[8] Other studies of birds have shown significant advances in the onset of breeding: by over eight days from 1971 to 1995 in 20 of 63 European bird species and by nine days from 1959 to 1991 for North American tree swallows.[9] In a third study, of 23 European pied flycatcher populations, there was a significant correlation between changes in the local spring temperature and the onset of egg-laying—the warmer the local temperature, the earlier the onset of egg-laying within the local population.[10]

A second category of climate change effects are shifts in the range of a species, with movement in the long term expected to be toward each pole and to higher altitudes.[11] Once again, a growing body of scientific evidence supports this. Near Antarctica, for instance, data indicate that several species of penguins, both sea ice–dependent and ocean-going species, have moved southward toward the pole.[12] Among more temperate birds, data indicate that 12 species in the United Kingdom have moved northward by, on average, 18.9 kilometers over 20 years.[13] Among insects, 23 species of dragonflies and damselflies in the United Kingdom expanded their ranges northward by on average 88 kilometers between 1960 and 1995.[14]

Cases of elevational shifts are also well documented. Among 16 Spanish butterflies, the lower elevational range has risen by, on average, 212 meters over 30 years.[15] In mammals, 7 of the 25 populations of the pika in the western United States have gone extinct since being recorded in the 1930s, but the populations that disappeared were at significantly lower elevations than those that survived.[16]

These numerous studies indicate quite clearly that climate change has had a significant

Table 1. Impact of Global Climate Change on Biodiversity over the Last Century

Ecosystem type	Impact
Forest	
Boreal	Low
Temperate	Low
Tropical	Low
Dryland	
Temperate grassland	Low
Mediterranean	Low
Tropical grassland and savanna	Moderate
Desert	Moderate
Inland water	Low
Coastal	Moderate
Marine	Low
Island	Low
Mountain	Moderate
Polar	High

Source: Millennium Ecosystem Assessment.

effect on terrestrial biodiversity, causing significant changes in a number of organisms across a wide array of ecosystems. Will these changes continue? And what will be their long-term effects? The IPCC and MA reports address the first of these questions. The IPCC concluded that it is "virtually certain" that recent warming trends will continue, and the MA projected that the impacts of climate change on biodiversity across all ecosystems will increase very rapidly.[17]

Answers to the second question are less clear. In some cases, the economic effects of climate change are likely to be positive; longer growing seasons and range shifts for agricultural crops benefit farmers in certain geographic regions, for example. Yet many of the impacts will be negative, including the direct and indirect effects of species relationships being disrupted and direct species extinctions.[18] One study that synthesized the changes of 11 multispecies interactions found that more than 60 percent of the interactions had been disrupted and become less synchronous over time due to the different responses of individual species to climate change, producing, in some cases, sig-

nificant negative consequences.[19] Habitat loss and other climate change effects can also result in significant population declines. Polar bears, for instance, are dealing with decreases in the extent and thickness of sea ice, resulting in both shrinking population sizes and reductions in mean body weight.[20] Thus climate change can significantly increase the probability of species extinction.

These individual accounts clearly demonstrate specific effects that climate change is having on terrestrial biodiversity. The larger scope of the problem is better illustrated by the IPCC, however. In an assessment of 29,000 observed changes in terrestrial biological systems, more than 89 percent of the significant changes were consistent with the direction of change expected as a response to global warming.[21] The IPCC concluded "with high confidence that anthropogenic warming over the last three decades has had a discernible influence on many physical and biological systems."[22] Some of these changes are likely to have, at least in the short term, positive economic and biological impacts. But many of the long-term impacts will undoubtedly be negative. Among the most significant of those are a 5 out of 10 chance of an increased extinction risk for 20–30 percent of plants and animals and an 8 out of 10 chance of major changes in ecosystem structure and function.[23]

With increased understanding of the long-term consequences of climate change and the greater probability of negative biological outcomes should climate change continue at its current rate, the need to grapple with the challenges of global climate change also increases. The IPCC noted that a wide array of technological, behavioral, managerial, and policy options are currently available; however, they also noted that more extensive action is required in order to reduce human vulnerability to future climate change.[24] As Tom Lovejoy of the Heinz Center for Science, Economics, and the Environment puts it, "Life on Earth is sending an urgent warning signal that climate change needs to be engaged with—and with an urgency and scale hitherto not contemplated."[25]

Threats to Species Accelerate

Elroy Bos

According to IUCN–The World Conservation Union, in its latest assessment of the state of life on our planet, the number of known threatened species reached 16,118 in 2006.[1] The ranks of those already facing extinction were joined by familiar species like the hippopotamus and desert gazelles, along with ocean sharks, freshwater fish, and Mediterranean flowers.[2] By now, 784 species on Earth have been declared extinct, and a further 65 are found only in captivity or cultivation.[3]

Of the 40,168 species assessed using the IUCN Red List criteria, one in three amphibians, a quarter of the world's coniferous trees, one in eight birds, and nearly one in four mammals are now known to be in jeopardy.[4] (See Table 1.) The term "threatened" includes three Red List categories of escalating threat: vulnerable, endangered, and critically endangered.

These numbers from the *2006 IUCN Red List of Threatened Species* demonstrate the ongoing decline of global biodiversity and the impact that humankind is having on life on Earth. Additions to the list in 2006 included a particularly familiar face—the polar bear. This charismatic mammal is now classified as vulnerable, as it is set to become one of the most notable casualties of Earth's rising temperature. The impact of climate change is increasingly felt in polar regions, where summer sea ice is expected to decrease by 50–100 percent over the next 50–100 years.[5] Polar bears are predicted to suffer more than a 30-percent population decline in the next 45 years as the ice floes they depend on when they hunt seals slowly disappear.[6]

LINKS pp. 42, 94, 98

Escalating threats to desert wildlife are unregulated hunting and habitat degradation. The dama gazelle of the Sahara, which was listed as endangered in 2004, has suffered an 80-percent crash in numbers over the past 10 years because of uncontrolled hunting and is now deemed critically endangered.[7] Other Saharan gazelle species are also threatened and seem destined to suffer the fate of the scimitar-horned oryx: extinct in the wild.[8]

In 2006, the Red List also included comprehensive regional assessments of selected marine groups. Sharks and rays are among the first such groups to be systematically assessed; of the 547 species evaluated so far, 20 percent are threatened with extinction.[9] This confirms suspicions that these mainly slow-growing species are extremely susceptible to overfishing and are disappearing at an unprecedented rate.

The plight of the angel shark and common skate, once familiar sights in European fish markets, illustrates dramatically the recent rapid deterioration of many sharks and rays. They have all but disappeared from sale.[10] The angel shark (moved from vulnerable to critically endangered) has been declared extinct in the North Sea, and the common skate (moved from endangered to critically endangered) is now very scarce in the Irish Sea and the southern North Sea.[11]

Freshwater species are not faring much better. They have suffered some of the most dramatic declines: 56 percent of the 252 endemic freshwater Mediterranean fish are threatened with extinction, the highest proportion in any regional freshwater fish assessment so far.[12] Seven species, including carp relatives *Alburnus akili* in Turkey and *Telestes ukliva* from Croatia, are now extinct.[13] Of the 564 dragonfly and damselfly species so far assessed, nearly one in three are threatened, including nearly 40 percent of endemic Sri Lankan dragonflies.[14]

Larger freshwater species, such as the common hippopotamus, are also in difficulty. One of Africa's best known aquatic icons, it has been listed as threatened for the first time and is classified as vulnerable, primarily because of a catastrophic decline in the number of hippos in the Democratic Republic of the Congo.[15] In 1994 this country had the second largest hippo population in Africa—30,000 after Zambia's 40,000—but today numbers have plummeted by 95 percent due to unregulated hunting for meat and the ivory in hippo teeth.[16]

The *IUCN Red List of Threatened Species* has become an increasingly powerful tool for conservation planning, management, monitoring, and decision-making. It is used by government agencies and nongovernmental organizations in at least 57 countries to compile national Red

Lists and is a focus for conservation action.[17]

Against the catalogue of decline, the latest data also show that conservation action does work. Following significant recoveries in many European countries, the numbers of white-tailed eagles doubled in the 1990s, and this species has been moved from the near threatened category to of least concern.[18] Enforcement of legislation to protect the species from being killed and measures to address threats from habitat changes and pollution have resulted in increasing populations.[19]

On Australia's Christmas Island, the seabird Abbott's booby was declining due to habitat clearance and an introduced invasive alien species, the yellow crazy ant, which had a major impact on the island's ecology.[20] The booby, listed as critically endangered in 2004, is recovering thanks to conservation measures and has now been moved to the endangered category.[21]

Other plants and animals highlighted in previous Red List announcements as under threat are now the focus of concerted conservation actions, which it is hoped will improve their conservation status in the near future. Some noteworthy examples are the 300-kilo-gram Mekong catfish of Southeast Asia, the Indian vulture, the humphead wrasse, and the Saiga antelope.[22]

These examples also illustrate a valuable lesson: bringing about the recovery of species on the edge of extinction is much more difficult and costly than preventing the decline in the first place by, for example, protecting habitat. They also underline the need for reliable scientific data on the status of species to guide recovery efforts and for quicker responses by governments and civil society when species or habitats come under threat.

Table 1. Threatened Species, by Major Groups of Organism, 2006

Organism	Described Species	Species Evaluated	Threatened Species	Threatened Species as Share of Known Species	Threatened Species as Share of Species Evaluated
	(number)	(number)	(number)	(percent)	(percent)
Vertebrates					
Mammals	5,416	4,856	1,093	20	23
Birds	9,934	9,934	1,206	12	12
Reptiles	8,240	664	341	4	51
Amphibians	5,918	5,918	1,811	31	31
Fishes	29,300	2,914	1,173	4	40
Subtotal	58,808	24,284	5,624	10	23
Invertebrates					
Insects	950,000	1,192	623	0.07	52
Mollusks	70,000	2,163	975	1.39	45
Crustaceans	40,000	537	459	1.15	85
Others	130,200	86	44	0.03	51
Subtotal	1,190,200	3,978	2,101	0.18	53
Plants					
Mosses	15,000	93	80	0.53	86
Ferns and allies	13,025	212	139	1	66
Gymnosperms	980	908	306	31	34
Dicotyledons	199,350	9,538	7,086	4	74
Monocotyledons	59,300	1,150	779	1	68
Subtotal	287,655	11,901	8,390	3	70
Others					
Lichens	10,000	2	2	0.02	100
Mushrooms	16,000	1	1	0.01	100
Subtotal	26,000	3	3	0.01	100
Total	1,562,663	40,168	16,118	1	40

Source: IUCN.

Invasive Species Drive Biodiversity Loss   S. Pagad and M. Browne

In 2005, the Millennium Ecosystem Assessment (MA) determined that "across the range of biodiversity measures, current rates of loss exceed those of the historical past by several orders of magnitude and show no indication of slowing."[1] Current trends in biodiversity loss show no indication of a slowdown. The MA lists invasive species as one of five direct drivers behind biodiversity loss (the others are land use change, climate change, overexploitation, and pollution).[2]

Only a small proportion of invasive alien species—living organisms that are moved around the world through human activity and global trade—actually cause harm. But this subset of introduced non-native species, whether brought in intentionally or unintentionally, has major ecological and socioeconomic impacts. And they are found in all major taxonomic groups.[3] (See Table 1.)

Invasive species cause a reduction in native biodiversity through predation, parasitism, hybridization, or competition with native species for habitats and resources.[4] They alter ecosystem functioning by causing changes in the nutrient and hydrological regime.[5] Socioeconomic damages can include loss of livelihoods and the expenditure of vast amounts of resources on control and mitigation of the risks caused by invasives.[6]

 LINKS pp. 94, 96

Nearly 30 percent of globally threatened birds are under threat from invasive aliens.[7] The problem is more severe on islands: 67 percent of this group of birds on islands are threatened by non-native species.[8] The extinction of at least 65 species of birds has been tied to predation by introduced rats, cats, pigs, dogs, and mongooses; to habitat destruction by sheep, goats, and rabbits; and to diseases caused by introduced pathogens.[9] For example, predation by rats has caused the near extinction of the Campbell Island teal in New Zealand, while avian malaria has caused the near extinction of birds in Hawaii.[10]

Candleberry myrtle or firebush, an invader of wet and mesic forests in Hawaii, forms dense, monotypic stands and has a negative effect on the recruitment and persistence of native plant species.[11] Firebush, a nitrogen-fixer, has altered primary successional ecosystems in the Hawai'i Volcanoes National Park by quadrupling inputs of nitrogen and is now reported to be spreading through drier submontane forests.[12]

Five major aquatic weeds that have spread over large areas of the natural and seminatural freshwater ecosystems of South Africa cause water availability and use problems.[13] They have reduced the quality of drinking water, increased the incidence of waterborne, water-based, and water-related diseases, and caused a decline in aquatic biodiversity.[14]

The global footprint of invasive alien species on biological diversity is yet to be quantified; a measure of the footprint will provide a better understanding of the need and priorities for effective conservation responses.

In 1993, the Office of Technology Assessment of the U.S. Congress documented economic damages of up to $97 billion between 1906 and 1991 due to 79 non-native invasive species.[15] More recently, David Pimentel and his colleagues at Cornell University estimated economic damages for the United States, the United Kingdom, Australia, India, South Africa, and Brazil to be in excess of $336 billion per year.[16]

Practical responses to biological invasions include preventing the intentional and unintentional introduction of invasive aliens, management and control of the ones already present and established, and mitigation of the risks and impacts they cause. The collection and exchange of authoritative data and information is a key component of these responses, and the wide dissemination of summary information helps raise public awareness. Examples of global, regional, national, and thematic information systems include the Global Invasive Species Database (GISD), the North European and Baltic Network on Invasive Alien Species, Pacific Island Ecosystems at Risk, and Non-indigenous Aquatic Species.[17] A network that will link all these information systems together, the Global Invasive Species Information Network, is also being developed.[18]

The most detailed and accurate data on inva-

Table 1. Examples of Some of the Worst Invasive Alien Species

Type	Examples
Microorganism	avian malaria
	banana bunchy top virus
Macro-fungi	crayfish plague
	Dutch elm disease
	frog chytrid fungus
Aquatic plant	caulerpa seaweed
	common cord-grass
	wakame seaweed
Land plant	salt cedar
	mile-a-minute weed
	Brazilian pepper tree
Aquatic invertebrate	Mediterranean mussel
	Northern Pacific seastar
	zebra mussel
Land invertebrate	Argentine ant
	Asian longhorned beetle
	Asian tiger mosquito
Amphibian	cane toad
	Caribbean tree frog
Fish	common carp
	large-mouth black bass
	Mozambique tilapia
Bird	Indian myna bird
	red-vented bulbul
Reptile	brown tree snake
	red-eared slider
Mammal	feral goat
	European rabbit
	ship rat

Source: Global Invasive Species Database.

initiative on invasive species led by the Global Invasive Species Programme.[20]

GISD profiles include information on the ecology, impacts, distribution, and range expansion of invasive alien species, along with images and descriptions, information about effective prevention and management options, and contact details for experts on each species. Users include natural resource managers, extension agents, environment and biodiversity specialists, quarantine and border control personnel, educators and students, and other individuals and organizations concerned with the environment.

The GISD has recently launched two new initiatives: the Global Register of Invasive Species (GRIS) and the Global Management Project Register (GMPR).[21] The GRIS will identify species with a history of being invasive by integrating invasive alien species checklist data generated by collection and observation databanks around the world. The GMPR will have case studies about prevention, eradication, control, and containment and mitigation activities.

Fortunately, those working on invasive species exhibit a willingness to share information and knowledge because they understand its importance for improving biodiversity outcomes. The GISD is just one of many responses to the need to collect and disseminate accurate, up-to-date, relevant information about invasive species. As awareness grows, people and communities are able to make informed choices that will have lasting effects on their descendants.

sive alien species at the global scale is available in the Global Invasive Species Database.[19] The GISD is a free searchable source of authoritative information about species that have a negative impact on biodiversity. It aims to facilitate effective prevention and management activities by disseminating specialist knowledge and experience to a broad global audience. Development of the GISD began in 1998 as part of the global

Ocean Pollution Worsens and Spreads

Brian Halweil

At the same time that marine scientists are reporting that the world's growing appetite for seafood may drive major fish populations to extinction in coming decades, humans are undermining marine health by using the oceans as a dumping ground.[1] (See Figure 1.) From inland farms and coastal sewage systems to trans-oceanic cruises and greenhouse gases, 80 percent of pollutants in oceans originate on land.[2] And although certain national and international laws have curbed oil spills and dumping by cruise ships, the amount of contaminants accumulating in oceans grows, even as the oceans' ability to dilute these substances declines.[3]

Many substances that are carried into the world's rivers and streams eventually find their way into coastal waterways and oceans. Around 60 percent of the wastewater discharged into the Caspian Sea is untreated, for example, while in Latin America and the Caribbean the figure is close to 80 percent and in large parts of Africa and the Indo-Pacific region the proportion is as high as 90 percent.[4] In the United States alone, more than 3.2 trillion liters of sewage—including human waste, detergents, and household chemicals—gush untreated into waterways every year.[5] Worldwide, an estimated $56 billion is needed annually to address this enormous wastewater problem.[6] By some estimates, the fastest-growing source of ocean pollution is the chemicals, human waste, and trash that run off of coastal city streets into ocean-bound storm drains.[7]

LINKS p. 26

More than half of the world lives in coastal areas (within 200 kilometers of shore) that cover just 10 percent of Earth's surface.[8] These coastal populations are increasing at twice the rate of inland ones.[9] An estimated 70 percent of the world's tropical coasts have been developed for housing, fish farms, or industrial ports, and the United Nations expects 90 percent will be developed by 2032.[10]

Nitrogen, phosphorus, and other nutrients from fertilizers, large livestock farms, and septic systems provoke explosive blooms of tiny plants known as phytoplankton, which die and sink to the bottom and then are eaten by bacteria that use up the oxygen in the water.[11] This oxygen starvation creates "dead zones" that make it difficult for fish, oysters, sea grass beds, and other marine creatures to survive.[12] There are now about 200 of these zones around the world, roughly one third more than just two years ago.[13] The most severe cases exceed 20,000 square kilometers, as in the Gulf of Mexico, the Bay of Bengal, the Arabian Sea, the East China Sea, and the Baltic Sea.[14]

The U.N. Environment Programme (UNEP) estimates that 46,000 pieces of plastic litter—including bits of packaging, cigarette lighters, plastic bags, and diapers—are floating on every square mile of the oceans, a figure that has increased threefold since the 1960s.[15] Marine conservation groups estimate that more than a million seabirds and 100,000 mammals and sea turtles die globally each year by getting tangled in or ingesting plastics.[16]

Even pollution of the atmosphere—in the form of greenhouse gases and resulting climate change—is taking a growing toll on ocean life. Climate change is altering fish migration routes, pushing up sea levels, leading to more coastal erosion, raising ocean acidity levels to a point where they threaten calcium-building species like corals and shellfish, and interfering with ocean currents that move vital nutrients upward from the deep sea.[17] The latter generates chaos among plankton, the foundation of the ocean food chain, that ironically help store carbon dioxide in the ocean floor as they die and decompose; the oceans have absorbed about half of the carbon dioxide produced by humans in the last 200 years.[18]

Humans suffer, of course, when ocean pollution reduces fish populations or stains the pristine nature of beach recreation. Industrial pollutants, like mercury or PCBs, that end up in water bodies are absorbed by fish we eat. In the last half-century, scientists around the world have tracked a 10-fold increase in pollution-fed algae blooms, which have produced toxins that poison sea life, seafood, and even humans swimming in and living near the ocean.[19]

Some of the most pernicious forms of ocean pollution are generated by the very people and industries that benefit directly from the pristine

Figure 1. Selected Ocean Pollution Hotspots

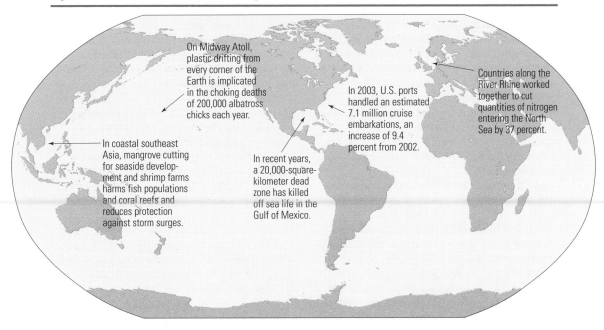

nature of the seas. Lax state and federal anti-pollution laws allow the world's growing fleet of more than 200 cruise ships to dump into the ocean untreated sewage from sinks and showers and inadequately treated sewage from toilets.[20] Once ships are three miles from shore, they can dump all untreated sewage, including bacteria, pathogens, detergents, and heavy metals. Each day, a standard cruise ship generates some 114,000 liters of sewage from toilets; 852,000 liters of sewage from sinks, galleys, and showers; seven tons of garbage and solid waste; 57 liters of toxic chemicals; and 26,500 liters of oily bilge water.[21]

Oceana, an international ocean protection group, launched a campaign to introduce Clean Cruise Ship legislation in California and at the U.S. federal level to prohibit dumping of boat sewage.[22] And following an aggressive 11-month grassroots campaign aimed at the world's second largest cruise line, Royal Caribbean agreed to install advanced wastewater treatment technology on all 29 of its ships.[23]

Since its formation in 1995, UNEP's Global

Programme of Action for the Protection of the Marine Environment from Land-Based Sources has helped reduce oil discharges and spills into the oceans by 63 percent compared with levels in the mid-1980s.[24] And tanker accidents have dropped by 75 percent, partly as a result of the shift to double-hulled tankers.[25]

The general public, perhaps because people care about eating seafood or because the oceans seem worth protecting, is also beginning to clean up pollution. Beginning in 1986, the Ocean Conservancy organized shoreline cleanups each fall.[26] To date, 6.2 million volunteers in International Coastal Cleanups have removed 49 million kilograms of debris from nearly 288,000 kilometers of coasts in 127 nations.[27] Nearly 60 percent of all debris is from recreational activities, including fishing lines and nets, beach toys, and food wrappers. An additional 29 percent is cigarette butts and filters.[28]

Bottled water—a general term referring to natural mineral water, spring water, and purified water supplied to consumers in bottles—is the world's fastest-growing commercial beverage. Global consumption of bottled water more than doubled between 1997 and 2005, reaching a total of 164.5 billion liters, or 25.5 liters per person.[1] (See Table 1.) While Europe and North America still dominate the bottled water market, consumption in Asia and South America has increased dramatically over the past five years, expanding at 14 percent and 8 percent a year respectively.[2]

The United States is the world's largest consumer of bottled water, with Americans drinking 28.7 billion liters in 2005.[3] But consumption per person is a different story: in 2005 each Italian, on average, drank more bottled water than anyone else in the world—192 liters, compared with 99 liters for Americans.[4] Among the top 10 countries, Brazil, China, and India have doubled or even tripled consumption between 2000 and 2005, though per capita intake in China and India is still far below the global average.[5] Altogether, almost three quarters of the world's bottled water is consumed in the top 10 countries.[6]

Worldwide, people buy bottled water in order to have safe drinking water, especially consumers in developing countries who face unreliable municipal water supplies, water scarcity, and continual water contamination.[7] In most industrial countries, however, where municipal water is better regulated, people drink bottled water also for better taste, for convenience, and as a substitute for other beverages.[8] In the United States, calorie-free bottled water has attracted consumers concerned about obesity.[9]

Urbanization, improved living standards, office working environments, and aggressive marketing strategies have helped boost the global sales of bottled water.[10] Home and office delivery of bottled water has become a popular service and supplies nearly 28 percent of the water consumed.[11]

The difference in cost between bottled and tap water is staggering: the bottled version costs from 240 times to more than 10,000 times as much.[12] The Pacific Institute, a California-based think tank, found that bottled water sold in most industrial countries costs $500–1,000 per cubic meter, compared with 50¢ per cubic meter of California's high-quality tap water.[13] Most of what consumers pay goes into production, packaging, transportation, advertising, retailing, marketing, and profits—not the water itself. In 2005, selling bottled water in the United States generated more than $10 billion in revenue.[14]

Social injustice remains a big concern in terms of bottled water consumption. People who desperately need a better supply of drinking water are usually not able to afford the bottled version.[15] In India, upper-class to lower-middle-class families are the main consumers, while tourists dominate bottled water consumption in rural areas.[16] The U.N. Development Programme's *Human Development Report 2006* notes that bottled water consumption generates nontangible health benefits but expands the gap between industrial and developing countries.[17]

Bottled water is regulated as a food product in the United States and Canada, while the European Union has two directives: one on natural mineral water and another on drinking water that includes bottled spring or purified water.[18] Regulation codes for bottled water generally cover the composition, contaminants, processing requirements, and labeling.[19] The Codex Alimentarius—an international food code initiated by the World Health Organization and the Food and Agriculture Organization—can be adopted by countries that lack national regulations.[20]

Based on a four-year study of the bottled water industry in the United States, including a test of more than 1,000 bottles of 103 brands of water, the Natural Resources Defense Council reported in 1999 that bottled water is not always safe to drink or better than tap water.[21] Regulations concerning bottled water are generally the same as tap water but weaker in certain standards for microbial contaminants. The U.S. Food and Drug Administration (FDA), which regulates bottled water at the federal level, permits this product to contain certain levels of fecal coliforms, while the Environmental Protection Administration does not allow fecal col-

Table 1. Consumption of Bottled Water, Total and Top 10 Countries, 2000 and 2005

Country	2000 Total Volume	2000 Share of Global Consumption	2000 Consumption Per Person	2005 Total Volume	2005 Share of Global Consumption	2005 Consumption Per Person
	(million liters)	(percent)	(liters)	(million liters)	(percent)	(liters)
United States	17,955	16.5	61.6	28,651	17.4	99.2
Mexico	12,464	11.5	124.3	18,861	11.5	179.7
China	6,012	5.5	4.7*	12,901	7.8	9.9*
Brazil	6,838	6.3	39.0*	12,252	7.4	65.8*
Italy	9,251	8.5	160.4	11,145	6.8	191.9
Germany	8,427	7.8	101.8	10,581	6.4	128.4
France	7,456	6.9	126.2	8,424	5.1	139.1
Indonesia	4,314	4.0	20.2*	7,633	4.6	33.3*
Spain	4,221	3.9	105.6	5,923	3.6	147.1
India	2,157	2.0	2.1*	6,177	3.8	5.6*
All others	29,391	27.1		41,982	25.5	
Total	108,517			164,530		
Global average			17.9			25.5

Source: International Bottled Water Association.
*These numbers are not directly available; they were calculated with population data from the U.S. Bureau of the Census.

iforms in city tap water.[22] And when violating the weaker FDA standards, bottled water may still be sold if it is labeled "containing excessive chemical substances" or "excessive bacteria."[23] Bottled water violations are not always reported to the public, or the products are recalled up to 15 months after the problematic water was produced, distributed, and sold.[24]

The environmental impacts of bottled water also need to be considered. Excessive withdrawal of natural mineral water or spring water to produce bottled water has threatened local streams and groundwater aquifers.[25] And producing, bottling, packaging, storing, and shipping bottled water uses significant amounts of energy.[26] In addition, millions of tons of oil-derived plastics—mostly polyethylene terephthalate (PET)—are used to make the water bottles.[27]

PET bottles have comparatively lower environmental impacts than glass or aluminum by requiring less energy to recycle or remanufacture, and they do not release chlorine into the atmosphere when incinerated, which PVC does.[28] But without proper recycling, massive amounts of PET bottles in the waste stream pose serious challenges to land uses as well as to water and air quality around landfills.[29]

In the United States, about 2 million tons of PET bottles end up in landfills each year.[30] According to the National Association for PET Container Resources, U.S. use of PET for bottled water without carbonization grew more than 20 percent in 2005, while usage for carbonated soft drinks dropped.[31] The recycling rate of PET rose slightly to 23.1 percent in the United States that same year, with a total of 2.3 million tons of waste generated. But this was still far below the 39.7-percent recycling rate achieved 10 years earlier.[32] Sales of plastic water bottles under 1 gallon have skyrocketed over the past decade in the United States, from 2.7 billion in 1997 to 28.6 billion in 2005.[33] Most of the water is consumed far from residence-based recycling programs. Adding a refund value—a nickel or dime—to the price of bottled water might give consumers an incentive to recycle. The 11 states embracing "bottle bills" with refund provisions have achieved three to four times the recycling rate of other states.[34]

Sustainable Communities Become More Popular E. Assadourian

In numerous communities around the world, people are working to reduce their impacts on the local as well as the global environment. Some are retrofitting existing communities, others are building new ones, still others are creating new programs in existing communities.

The growing global ecovillage movement is one of the more developed examples of this trend. An ecovillage, according to one widely accepted definition, is a "human scale full-featured settlement in which human activities are harmlessly integrated into the natural world in a way that is supportive of healthy human development and can be successfully continued into the indefinite future."[1] So far, these rather stringent criteria provide an ideal that ecovillages strive for rather than a standard actually achieved.

According to the Global Ecovillage Network directory, there are currently 375 ecovillages around the world.[2] (See Table 1.) While all ecovillages strive toward a similar goal, the diversity found among them is striking. They can be found in rural, suburban, and urban areas and in industrial as well as developing countries.[3] This figure does not reflect the total number of communities striving to be sustainable, however; it excludes, for example, cohousing communities and several broader networks of sustainable villages.[4]

In the mountains outside of Asheville, North Carolina, there is a rural ecovillage of 60 individuals.[5] Started 13 years ago, it is designed to grow to 160 once finished.[6] Homes there are built mainly from wood harvested from the local forests, water comes from mountain springs and rainwater harvesting, and electricity is generated from solar photovoltaic cells and a microhydro generator.[7]

p. 54

Another rural ecovillage, Mbam, is located in the Siné-Saloum delta in Senegal.[8] Along with using low-impact and appropriate technologies such as solar ovens and permaculture, one of its primary activities is restoring the health of mangrove forests to help protect the coast from salinization.[9]

A suburban ecovillage in Denmark, Munksøgård, is about a half-hour train ride from Copenhagen.[10] Some 230 residents live in 100 apartments clustered in five groups.[11] Munksøgård maintains a 24-hectare organic farm that provides food for the community.[12] It is the largest ecological building project in the country and in 2000 received first place in a Danish competition for the best sustainable design for the twenty-first century.[13]

Ecovillages are also being established in urban areas. In Porto Alegre, Brazil, for instance, a community for 28 families was built in 2002.[14] The group used sustainable building methods and materials (such as passive solar design and locally sourced materials) and included gardens, grass roofs, and artificial wetlands to process sewage.[15] Along with serving as a demonstration project for affordable, sustainable housing, the community—through a consultancy firm it established—is helping to start two more ecovillages in the city.[16]

Many ecovillages have made great strides in reducing their ecological impact. A recent analysis found that the ecological footprint per capita at Findhorn, an ecovillage in Scotland, was about 60 percent of the average footprint in the United Kingdom.[17] And in the Sieben Linden ecovillage in Germany, per capita carbon dioxide emissions were just 28 percent of the German average.[18]

Beyond ecovillages, a much broader set of communities is also providing lessons in sustainable living. Certain religious communities have chosen to lead simple lives, even when modern technologies are readily available. In the United States, for example, some Amish communities do not use electricity or motors (although most Amish do not ban the use of motors) and thus have much smaller impacts on the global environment.[19] Many homesteading communities, in which the majority of residents sustain themselves with farming and use more local resources, have much smaller environmental impacts than other communities.[20]

Yet most people raised in the consumer society have no interest in "going back to the land." But there are many more mainstream opportunities to reduce the environmental impacts of

Table 1: Ecovillages, by Region

Region	Number
Europe	137
North America	107
Latin America	58
Asia/Oceania	52
Africa/Middle East	21
Total	375

Source: Global Ecovillage Network.

daily life at the community level—some of which do so more as a byproduct of trying to rebuild social connections in a culture where social ties are diminishing.[21]

The cohousing movement, for example, is primarily focused on improving the quality of life of community members by designing housing that facilitates social ties.[22] Cohousing efforts involve a shared community building, which means individual homes can be smaller; a clustered housing pattern, which allows more of the community's land to be preserved in a natural state (if in a rural area); occasional shared meals; and some shared services and major appliances (such as cars, power tools, and other major pieces of equipment that are used infrequently).[23] This tends to make cohousing communities more sustainable than the average community.[24] While exact numbers of cohousing communities are difficult to find, an estimated 229 of them are found in North America and more than 250 in Europe, with the majority of these located in Denmark—the birthplace of cohousing.[25]

Mainstream developers are also starting to incorporate sustainability into their designs for new housing. Peabody Trust, which provides affordable housing for more than 50,000 people in London, created an 82-unit housing complex called the Beddington Zero Energy Development (BedZED).[26] As the name suggests, the goal of the community is to produce as much energy as it uses, which it strives for through a combination of passive solar design, energy efficiency measures, a community-scale power plant that provides electricity and hot water and is fueled by wood waste, and greater use of walking, cycling, and public transit.[27] A resident living at BedZED has just 60 percent of the ecological footprint of an average individual in the United Kingdom.[28]

International agencies, too, are helping to support community-initiated sustainable development efforts. The Global Environment Facility's COMPACT program (Community Management of Protected Areas Conservation), for instance, provides grants of less than $50,000 to communities in World Heritage Sites such as Mount Kenya to help villages create projects that improve people's lives while also reducing their impact on the surrounding ecosystems.[29]

With dramatic changes from a warming climate and the unsustainable use of many of the ecosystem services on which humans depend, more communities are trying to address sustainability issues.[30] Many are trying to localize farming, reduce energy use, and create stronger local businesses.[31] Already, communities have established local food co-ops, community-supported agriculture programs, carpools, and other ways to connect a community while lowering environmental impacts.[32]

Broader networks have sprung up around the world to spread these sustainable practices. The Relocalization Network, started in 2003, helps coordinate 148 local groups in 12 countries, providing an online forum for local communities trying to become more sustainable and less dependent on a fragile, globalized economic system.[33] And many ecovillages, such as The Farm in Tennessee, offer classes on how to increase sustainability at the community level.[34]

In Sri Lanka, the Sarvodaya Shramadana movement now works with 15,000 villages, helping them to develop economically in a more sustainable way.[35] The Sarvodayan "no poverty, no affluence" model is based on addressing basic needs such as access to food, shelter, clean water, and basic health care, but it considers nonmaterial needs like access to a clean and beautiful environment, a well-rounded education, and spiritual sustenance equally important.[36]

Social and Economic Features

A child earns money by crushing stones in order to help pay his school fees, Lusaka, Zambia.

© 2004 Abdel Mambwe Chibu, Courtesy of Photoshare

▶ Progress Toward the MDGs Is Mixed

▶ Literacy Improves Worldwide

▶ Child Labor Harms Many Young Lives

▶ Informal Economy Thrives in Cities

▶ Socially Responsible Investment Grows Rapidly

For data and analysis on these social and economic topics and on others such as nano-technology and corporate responsibility reporting, go to www.worldwatch.org/vsonline.

As the mid-point approaches in efforts to meet the U.N. Millennium Development Goals (MDGs) by 2015, progress toward achieving them is uneven. Although many countries have made important strides, greater efforts by both industrial and developing nations are needed to accelerate action.[1]

World leaders first adopted the MDGs in preliminary form at the U.N. Millennium Assembly in September 2000.[2] In addition to eight overarching goals, the MDGs have 18 specific targets, most of which are to be achieved by 2015. These include cutting poverty and hunger rates in half from their 1990 levels, reducing child mortality by two thirds, and halving the proportion of people lacking access to clean drinking water and adequate sanitation.[3] (See Box 1.) The United Nations also developed a set of 48 indicators to monitor progress toward the MDGs.[4]

Current estimates suggest that the world is on track to meet the overarching income poverty target—to halve by 2015 the proportion of people living on less than $1 per day.[5] The share of the developing world's population living in extreme poverty declined from 27.9 percent in 1990 to 19.4 percent in 2002 (the most recent year for which data are available).[6] This drop was driven in no small measure by particularly rapid progress in poverty reduction in much of Asia. Trends have unfortunately been less encouraging in other regions, particularly in sub-Saharan Africa, where the absolute number of people living in extreme poverty increased by 140 million between 1990 and 2002 due to population growth despite a modest decline in the overall poverty rate.[7] If present trends continue, few African countries are expected to meet the 2015 target for income poverty.[8]

The situation is even bleaker when it comes to hunger. Although the proportion of people suffering from hunger worldwide declined modestly over the last decade, the absolute numbers are rising, with an estimated 834 million chronically undernourished people in developing countries at last count.[9]

Child mortality trends are somewhat more

LINKS pp. 50, 110, 120, 122

positive, with 2.1 million fewer deaths among children under five in 2004 than in 1990.[10] Still, the U.N. Development Programme estimates that at current rates of progress, the target of reducing by two thirds the child mortality rate by 2015 will be missed by some 4.4 million deaths that year.[11]

Major challenges also remain in efforts to meet other human development goals, such as those related to gender equity, maternal health, and deadly diseases such as HIV/AIDS, malaria, and tuberculosis. The World Bank reports that all regions are off track on at least some of these goals and that South Asia and sub-Saharan Africa are off target on all of them.[12]

Nonetheless, there are also some encouraging signs. For example, the number of countries set to meet the goal of providing universal primary education by 2015 has increased significantly since 2000, and gender gaps in access to primary and secondary education are narrowing.[13] The number of AIDS patients in developing countries with access to treatment has increased rapidly, rising from less than 100,000 in 2000 to nearly 1 million in 2005.[14] Efforts to combat malaria by providing bednets and better treatment options are also expanding quickly: the distribution of insecticide-treated bednets increased 10-fold in sub-Saharan Africa between 1999 and 2003.[15]

Progress toward ensuring environmental sustainability is mixed at best. The MDGs call on countries to cut in half by 2015 the proportion of people without access to safe drinking water and adequate sanitation. Recent analyses suggest that the world is on track to meet the drinking water target: the share of people using drinking water from improved sources rose from 78 percent in 1990 to 83 percent in 2004.[16] But more than 1 billion people worldwide still lack access to an improved water supply, including two out of every five individuals in sub-Saharan Africa.[17] And as for sanitation, the picture is substantially worse. Although the share of people in the developing world with access to improved sanitation facilities increased from 49 percent in 1990 to 59 percent in 2004, 2.6 billion people still lack connections to public sewers or even access to simple pit latrines or other improved sanitation facilities.[18] Current estimates suggest that the overall 2015 target for sanitation is unlikely to be met.[19]

The MDGs also call on countries to integrate the principles of sustainable development into country policies and programs and to reverse the loss of environmental resources. At the 2002 World Summit on Sustainable Development in Johannesburg, governments adopted a number of related additional targets, including restoring fisheries to their maximum sustainable yields by 2015 and significantly reducing the rate of biological diversity loss by 2010.[20] But progress toward these goals has been inadequate. Jeffrey Sachs, who for several years was Special Advisor for the MDGs to U.N. Secretary-General Kofi Annan, recently noted that there was little awareness of the 2010 biological diversity target and that the goal was not being achieved.[21] He called the environment the biggest challenge facing humanity and noted that gains against poverty could be "washed away" by forces such as tropical storms, massive flooding, droughts, loss of snow melt, and desertification.[22]

Efforts to develop a global partnership for development have also been uneven. On the encouraging side, donor aid to developing countries has risen steadily since 1997, reaching $106 billion in 2005.[23] But aid expenditures continue to be unequally distributed, in part due to political calculations. More than 60 percent of the increase in official development assistance (ODA) between 2001 and 2004 went to just three countries—Afghanistan, the Democratic Republic of Congo, and Iraq—which between them are home to fewer than 3 percent of the developing world's impoverished people.[24] Furthermore, stepped-up debt relief has accounted for over half of the increase in ODA since 1997 and three quarters of it in 2005.[25] Although this has contributed to a steady reduction in debt service payments for 29 heavily indebted poor countries since 1998, there is no guarantee that it will continue or that governments will channel the savings into efforts to meet the MDGs.[26]

Box 1. Millennium Development Goals and Targets

1. Eradicate extreme poverty and hunger
By 2015, reduce by half both the proportion of people living on less than $1 a day and the share suffering from hunger.

2. Achieve universal primary education
Ensure that by 2015 all boys and girls complete a full course of primary schooling.

3. Promote gender equality and empower women
Eliminate gender disparity in primary and secondary education, preferably by 2005, and at all levels by 2015.

4. Reduce child mortality
By 2015, reduce by two thirds the mortality rate among children under five.

5. Improve maternal health
By 2015, reduce by three quarters the maternal mortality rate.

6. Combat HIV/AIDS, malaria, and other diseases
Halt and begin to reverse the spread of HIV/AIDS, malaria, and other major diseases by 2015.

7. Ensure environmental sustainability
Integrate the principles of sustainable development into country policies and programs and reverse the loss of environmental resources. By 2015, cut in half the proportion of people without access to safe drinking water and sanitation. By 2020, improve significantly the lives of 100 million slum dwellers.

8. Develop a global partnership for development
Develop an open trading and financial system that is rule-based and nondiscriminatory and that includes a commitment to good governance, development, and poverty reduction. Address the special needs of least developed countries, including through tariff- and quota-free market access, enhanced debt relief, and more generous development assistance for countries committed to poverty reduction. Address the special needs of small island developing states and landlocked countries. Make debt sustainable, increase youth employment, and provide access to essential drugs and new technologies.

Source: United Nations.

Alana Herro

In 2000–04, the portion of adults in the world with reading and writing skills reached 82 percent, up from 75 percent in 1990.[1] (See Table 1.) The net enrollment ratio of children of eligible age entering primary school was 87 percent, up from 81 percent in 1991.[2] Although data are unavailable for several countries in conflict or post-conflict situations, these trends indicate progress toward international goals set in 2000 of meeting the basic learning needs of people of all ages within a generation.[3]

Although the illiterate population has dropped from its 1990 level of 874 million, there are still 781 million people over the age of 14—or one in five adults—who lack basic literacy skills.[4] Some 98 percent of illiterate people live in the developing world.[5] The region with the lowest literacy rate is Africa, at 62.5 percent, but Asia has the largest illiterate population—some 546 million people.[6] In the poorest countries, only about half of all adults can pass basic literacy tests.[7]

Some developing countries have been making significant increases in adult literacy, however. Egypt's literacy rate jumped from just 47.1 percent in 1990 to 71.4 percent in 2000–04, while Ethiopia saw a 16.6-percent increase in the literacy rate and China registered a 52-percent drop in the absolute number of illiterates in the same period.[8]

The three regions furthest from universal primary education saw improvements in enrollment: the Arab States, for example, saw a 6-percent increase from 1999 to 2004.[9] In South and West Asia, enrollment during those five years grew by 19 percent, and in sub-Saharan Africa there was a 27-percent increase.[10]

LINKS pp. 50, 108, 112

Education is a powerful tool against poverty: it is linked to higher income levels.[11] And the better educated a person is the more likely he or she is to report being in good health, regardless of income.[12] Recent studies even show a positive correlation between life expectancy and the number of years of education a person has.[13] For these and other reasons, education is essential to sustainable development.[14]

Almost two thirds of the people in the world who lack literacy skills are female, and in any region of the world an illiterate adult is most likely to be a woman.[15] (See Table 2.) Central and Eastern Europe have the largest share of the illiterate population being female.[16] Globally, women appear to be stopping their education at lower and lower levels, a trend that could erode education gains.[17]

Great strides have been made toward gender parity in primary education, however. Of the 181 countries with 2004 data available, some two thirds have achieved gender parity in primary education.[18] And for every 100 boys in primary school, there are 94 girls.[19]

Efforts focused on women's education tend to increase female participation and earnings in the labor force and to allow for more effective transfer of the benefits of education—health, educational opportunities, and more—from one generation to the next.[20] On average, a child whose mother has no education is twice as likely to not be in school as a child with an educated mother.[21]

Investment in girls' education results in some of the greatest returns of all development investments.[22] It is linked to higher crop yields and per capita income increases, for instance, and with lower rates of HIV infection and infant mortality.[23] In addition, education for girls reduces fertility rates, as educated women are more likely to delay marriage and childbearing, use reliable family planning methods, and

Table 1. Adult Literacy Rate, 1950–2004

Year	Adult Literacy Rate
	(percent)
1950	55.7
1960	60.7
1970	63.4
1980	69.7
1990	75.4
2000–04*	81.9

* Rate combines data for most recent year available in each country.
Source: UNESCO.

Table 2. Male and Female Adult Illiteracy Rates, Total and in Industrial and Developing Countries, 2000–04

Region	Men	Women
	(percent)	
Industrial Countries	1	1
Developing Countries	17	30
World	13	23

Source: UNESCO.

have fewer and healthier babies.[24]

Developing countries have unique challenges to achieving universal education. In sub-Saharan Africa, for example, nearly 10 percent of children under 17 years of age have lost at least one parent to HIV/AIDS, and an orphan is 13 percent less likely to be in school than a non-orphan.[25] Some 80 percent of people with disabilities live in developing countries, and it is estimated that more than one third of out-of-school children have a disability.[26] Poor children are less likely to attend school: the number of children not attending school in the poorest 20 percent of households is more than triple that in the wealthiest 20 percent.[27] Child soldiers and other youngsters affected by conflict represent another sector facing acute obstacles to education, as do sexually exploited children and those who are pressured into the labor force.[28]

While funding for education is on the rise, it is estimated that $11 billion a year in development assistance is required to achieve the goal of Education For All (EFA) agreed to in 2000, which is more than twice the current level of aid for basic education.[29] Some countries, including Italy, Nicaragua, Saudi Arabia, and Tunisia, significantly increased the share of their gross domestic product spent on education between 1990 and 1998–2000.[30] Others, like Bulgaria, Canada, and Uruguay, reduced the percent allocated to this sector.[31]

Many developing countries have abolished school fees, leading to a surge in primary school students. After Kenya removed school fees in 2003, 1.2 million additional students entered the school system; in 2005 Burundi enacted the same policy and increased enrollment by 500,000.[32]

At least 160 countries committed to achieving six EFA goals, which include a special focus on Early Childhood Care and Education (ECCE).[33] ECCE, like education for girls, is a particularly cost-efficient means of approaching the goals of EFA.[34] Studies suggest ECCE is also exceptionally effective at offsetting disadvantage and inequality for poor and culturally excluded children.[35]

Enrollment in pre-primary education has nearly tripled since the middle of the 1970s, though coverage remains very low in most of the developing world.[36] Most regions now have nearly as many girls as boys enrolled in pre-primary education.[37]

Countries that have made significant progress toward universal education promote policies that allocate public funds to education adequately and equitably and that promote high enrollment, especially for girls.[38] Successful countries also have policies that give women the right to own property and the ability to earn an independent income.[39] ECCE programs are most effective in promoting Education For All goals when they are taught in the child's native language, challenge gender stereotypes, mainstream children with disabilities, and are combined with other services such as health care, nutrition, and social services.[40]

Child Labor Harms Many Young Lives

Zoë Chafe

The number of child laborers dropped by 27.8 million between 2000 and 2004, according to the International Labour Organization (ILO), a U.N. agency.[1] Citing this 11.3-percent decline, the ILO declared that the end of child labor is "within reach."[2] Despite the encouraging trend, nearly 218 million children worldwide were engaged in child labor in 2004, the most recent year with data.[3] (See Table 1.) The term "child labor" includes all children between the ages of 5 and 11 who are involved in economic activity, children aged 12–14 who perform more than a few hours of permitted light work a week, and children aged 15–17 who are engaged in hazardous work.[4]

In the first four years of this decade, children who are considered economically active—a broader category than child labor—fell by 34.5 million, or 9.8 percent.[5] (All children who report having worked at least one hour on any day during a seven-day period are considered economically active.)[6]

The percentage of children who are economically active varies greatly by region.[7] Sub-Saharan Africa has the highest rate, with more than one in every four children 5–14 years old at work.[8] In Latin America, the figure is just 5 percent.[9] Asia and the Pacific region is home to 122.3 million children who are economically active, nearly two thirds of the global total.[10]

LINKS pp. 44, 54, 108, 114

The ILO estimates that it would cost $760 million to eliminate child labor, or about $38 million a year over 20 years.[11] In return, the potential quantifiable benefits in improved education and health are estimated at $4 trillion—more than five times the investment.[12] Brazil is an example of a country that has quickly reduced child labor rates—the number of five-to nine-year-olds working fell by 60 percent in 12 years.[13] The ILO points to two Brazilian social programs—*bolsa escola* and *bolsa familia*—that provide low-income families with stipends to keep their children in school and that raised primary school attendance to 97 percent as key factors in the rapid reduction of child labor.[14]

In Liberia, many children of rubber-tapping families that work for Firestone Tire and Rubber Company do not go to school. They work beside their parents, up to 12 hours a day, to fulfill high quotas by carrying heavy buckets of pesticide-laden latex on their heads.[15] If they do not tap enough trees, meager family wages are halved.[16] A coalition of U.S. and Liberian organizations is pressuring Firestone to reform its labor practices.[17] Unfortunately, this is just one example of child labor used in products sold throughout the world.

For 10 years, reports have exposed the expansive use of child labor in the cocoa industry.[18] With an estimated 70 percent of the world's cocoa grown in West Africa, child slavery in this industry is integral to the ongoing political instability of countries like Côte d'Ivoire, in the same way that profits from "blood diamonds" were used to fund ongoing violence in Angola and Sierra Leone.[19]

Pressure from the U.S. Congress resulted in the Harken-Engel Protocol—a promise by the chocolate industry to voluntarily end child labor on cocoa farms by July 2005.[20] The deadline passed, but no significant progress is apparent.[21] The International Labor Rights Fund subsequently filed suit against cocoa buyers and importers Nestlé, Archer Daniel Midlands, and Cargill to represent children from Mali who were brought to Côte d'Ivoire and forced to work long hours for no pay, picking cocoa beans that the companies then purchased.[22] In August 2006, the companies filed a brief asserting that, as buyers, they cannot control the labor force used to pick cocoa beans.[23] As of March 2007, both sides were waiting to learn if the case would go forward.[24]

In late 2006, U.S. Secretary of Labor Elaine Chao announced a $4.3-million initiative to eliminate the "worst forms of child labor" within the cocoa industry.[25] Researchers working under this initiative will study the health of exploited children, train officials in Côte d'Ivoire and Ghana to monitor for child labor, and report on progress toward implementing a child-labor-free cocoa certification system in those two countries.[26]

The vast majority of child laborers (69 percent) work in agriculture.[27] Another 22 percent

Table 1. Child Laborers, Total and Those Doing Hazardous Labor, 2000 and 2004

Population	2000	2004
Total population of children	1,531 million	1,566 million
Child laborers		
Number	246 million	218 million
Proportion of all children	16.0 percent	13.9 percent
Children doing hazardous labor		
Number	171 million	126 million
Proportion of child laborers	69.5 percent	58.0 percent

Source: *International Labour Organization.*

work in retail, restaurants, or other service economies, and the remaining 9 percent do industrial work such as mining or manufacturing.[28] Agricultural work frequently entails long hours in hot environments, exposure to pesticides, heavy loads, and injury from sharp tools.[29] In Egypt, more than 1 million children work to manually remove pests from cotton plants.[30] And in the United States, an estimated 300,000 children are hired to weed and pick commercial crops.[31] According to the ILO, the number of children working in agriculture outnumbers sectors that have received more attention (such as carpet weaving or garment manufacturing) by a ratio of nearly 10 to 1.[32]

The ILO estimates that 1.2 million children are sold into labor each year, in transactions that total as much as $10 billion, and that about one sixth of this trade affects African children.[33] Though these children are subjected to horrific work conditions and brutality, their families—many of whom live on less than $1 per day—often argue that learning work skills is a more positive prospect than constantly trying to find sufficient food at home.[34] Traffickers may play to this sentiment when convincing families to sell their children.

Children also become vulnerable to hazardous labor and trafficking in the aftermath of natural disasters. Those whose parents have died

or become unable to work must suddenly fend for themselves. Also, disasters cause many families to become temporarily separated, enabling traffickers to capitalize on the ensuing chaos. If schools are damaged or teachers are unable to work, children may turn to hazardous work. With the number of natural disasters increasing, the post-disaster scenario is a major concern for children's rights advocates.[35]

To counteract this alarming trend, many African countries have recently passed anti-trafficking laws. Burkina Faso reports that the formation of village surveillance committees helped police find and free 644 children in 2003.[36] And in three years, the International Organization for Migration claims to have freed 587 children from the fishing industry in Ghana's Lake Volta region.[37]

Children are engaged in domestic labor as well. India, where as many as 15 million children have been sold into labor, extended its Child Labour Act in October 2006 and banned children younger than 14 from working as domestic servants, at tea stands or food stalls, in restaurants or hotels, or in the hospitality industry.[38] A BBC report filed two months later, however, found that police misunderstood the law or were reluctant to enforce it.[39] After three girls, ages 6 to 13, were rescued from jobs as domestic servants in Faridabad, a city just outside Delhi, police refused to prosecute the girls' employer, saying that the law only applied when the children were not being paid or had been trafficked, neither of which was true in this case.[40]

The Convention on the Rights of the Child is an international document that says children under 18 years of age have the right "to be protected from performing any work that is likely to be hazardous to the child's health or physical, mental, spiritual, moral, or social development."[41] Only two countries have not agreed to implement the rights spelled out in the convention: Somalia and the United States.[42]

Informal Economy Thrives in Cities

Alessandra Delgado

For many poor people in urban areas, the primary means of economic survival is the production or sale of goods or services through semi-legal or illegal ventures, known as the informal economy.[1] Conservatively, informal employment accounts for half to three quarters of all nonagricultural employment in developing countries: 48 percent in North Africa, 51 percent in Latin America, 65 percent in Asia, and 72 percent in sub-Saharan Africa.[2]

In the 13 principal metropolitan areas of Bogotá, Colombia, 58.5 percent of workers are classified as informal.[3] In Bolivia, the informal sector provides an estimated two thirds of the gross domestic product (GDP), and in Peru the figure is 58 percent.[4] Because of the sheer number of workers, clients, budgets, and transactions involved in informal markets, legality is marginal; informality is the norm.[5]

The greatest increase in the informal economy since 1990 has occurred in sub-Saharan Africa, Latin America, and Central Asia—often accounting for more than 50 percent of GDP.[6] (See Figure 1.) The last few years have seen a continuation of this trend, with Africa and Latin America having the highest levels of informality.[7] In contrast, in Europe the growth of informality is slowing and even declining in the wake of extensive microeconomic reforms, while in East Asia, where firms face smaller regulatory and tax burdens, the informal economy remains stable at fairly low levels.[8]

The weight of such markets becomes clear when considering that economic power is concentrated in the cities. The purchases made by urbanites, who cannot live off the land, form the foundation of national economies.[9] Although 60 percent of the labor force in India is in the agricultural sector, for instance, they produce only 20 percent of the GDP, while the 28 percent of the population working in services provide 61 percent.[10]

Several factors combine to create the unique yet common pattern of informal markets. First, exaggerated government intervention in civil society and economic activity often creates "hyper-bureaucratization" that deters citizens

pp. 54, 108, 112

from pursuing a legal path.[11] For example, it is virtually impossible for 90 percent of Tanzanians to enter the legal economy.[12] A poor entrepreneur who obeyed the law would, over 50 years of business life, pay $91,000 to the national government for licenses, permits, and approvals and spend 1,118 days in government offices petitioning for them.[13] A private company can only be incorporated in Dar es Salaam and would cost nearly $2,700—almost four times the average annual wage of an ordinary Tanzanian.[14] Similarly, in Peru the constant increase in sales taxes—which went from 5 to 15 percent from 1978 to 1987 and today stand at 19 percent—favored expansion of the informal sector.[15]

Second, governments lack the resources to meet the demands of urbanization and enforce laws. Heavily indebted governments with limited tax collection and with convoluted and uninformed bureaucracies cannot provide adequate social expenditures.[16] Rapid urbanization in developing countries has created pressures that have constrained the capacity of cities to provide adequate employment, waste disposal, water supply, food supplies, and housing.[17] Urbanization itself has thus bred new types of economic arrangements and social conditions.

Third, as businesses are unable to create jobs as fast as demand increases, people must find a way to survive outside of regulated employment.[18] And fourth, many national and international companies prefer informal employment relations that allow them to be flexible during production cycles and that reduce labor costs.[19]

Thus it is not uncommon to visit an emerging market and perceive chaotic and unregulated yet bustling economies. Dharavi in India, the largest and most established of Mumbai's slums, by one estimate houses up to 10,000 small factories, almost all of them illegal and unregulated.[20] The factories provide an income for the approximately 1 million people who live in an area barely half the size of New York City's Central Park.[21] Although the concentration of businesses could easily deter consumers, the large scale at which informality occurs yields an estimated $665 million in annual revenue.[22]

On a national scale, in Haiti untitled rural and urban real estate holdings are together worth some $5.2 billion—four times the assets of all the legally operating companies in Haiti, nine times the assets owned by the government, and 158 times the value of all foreign direct investment in Haiti up through 1995.[23]

As the economic potential is great, the economic loss is equally substantial. Workers and enterprises receive little if any legal protection or worker benefits, they are the target of bribery, and they often face competitive disadvantages in terms of larger formal firms in capital and product markets.[24] Variations in incomes are great: in Bolivia, the owner of a small informal business might have an average income 12 times the national minimum wage, while informally paid workers and domestic servants make around half the minimum wage.[25]

Furthermore, there are indirect costs to informality. The unsafe working conditions found in the unregulated businesses of Dharavi, India, for example, are common throughout the world. In dark unventilated foundries, workers ladle molten metal into a belt-buckle mold held between their bare feet.[26] In another warehouse, men smeared from head to toe in blue ink strip the casings from used ballpoint pens so they can be melted down and recycled; few wear gloves or other protective gear, despite exposure to solvents and other chemicals.[27] Environmental and health hazards are just one of the realities workers have to withstand to be able to produce with minimal resources.

The indirect costs also exact a hefty social price. Even though the informal market has local arrangements that help keep track of transactions, the legitimacy of these informal rights is still too locally politicized compared with those that are protected by national law.[28] The inability to determine the rightful owner of resources creates or exacerbates conflicts

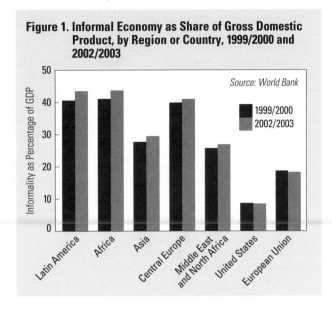

Figure 1. Informal Economy as Share of Gross Domestic Product, by Region or Country, 1999/2000 and 2002/2003

throughout the world.[29] In Bangalore, India, extortion exists even in hospitals: new mothers have their infants whisked away by an attendant who demands a bribe.[30] If you want to see your child, families are told, the price is $12 for a boy and $7 for a girl.[31] Such new "enterprises" are the result of a combination of a real need and a lack of regulation.

The lack of regulation distorts economic and social systems. With little or no unbiased and standardized regulation, most potential assets in emerging markets have not been identified or realized, there is little accessible capital, and economies are constrained and sluggish.[32] It is not surprising, then, that extensive preliminary research shows that countries with a sophisticated legal and political system and stronger protection of physical and intellectual property rights experience higher economic well-being.[33]

Socially Responsible Investment Grows Rapidly *Gary Gardner*

Socially responsible investment (SRI)—investment and advocacy designed to help promote sustainable economic activity—continues to grow rapidly in industrial countries and is beginning to emerge in a few developing nations as well. Because SRI is defined differently in different countries, a global figure for total SRI has not been established. But in every market in which this new breed of investment has a foothold, it is on the rise.

SRI volume is greatest in the United States, with a total value of $2.29 trillion in 2005.[1] (See Table 1.) Europe's commitment to socially responsible investing is rising rapidly, and in 2005 reached $1.22 trillion.[2] Canada, Australia, New Zealand, and Japan have smaller but growing SRI sectors.[3] And SRI funds have been established recently in Malaysia and Singapore.[4]

The share of total managed investments going into socially responsible economic activity is still small to modest, ranging from less than 1 percent in Japan to more than 9 percent in the United States.[5] But SRI growth has been rapid over the past decade, although national monitoring bodies measure different attributes of SRI, making growth rates difficult to compare. In Australia, SRI funds under management grew 36-fold between 2000 and 2006.[6] In the United States, they grew more than threefold between 1995 and 2005.[7] Canadian SRI grew nearly eightfold between 2004 and 2006.[8] And in Europe, SRI grew by some 36 percent between 2003 and 2006.[9]

LINKS p. 54

SRI primarily involves applying ethical screens to personal and institutional investments to ensure that funds are directed toward sustainable activities and away from unsustainable ones. Funds can use "negative" screens, meaning that they prohibit investment in companies or funds involved in specific activities such as tobacco production or nuclear power. "Positive" screens, a more recent SRI tool, encourage investments in companies that generate economic activity consistent with sustainability, such as solar power or microfinance. In Japan, where SRI is relatively new, almost all screens used are positive ones.[10]

In the United States and Europe, SRI activity also includes efforts to use shareholder power to steer corporate behavior in ethical directions. The number of resolutions on social and environmental issues introduced at annual shareholder meetings in the United States grew 16 percent between 2003 and 2005.[11] The number with enough support to reach a vote increased by 22 percent in the same period.[12] In 2006, for example, some Wal-Mart shareholders put forth a resolution asking the company to produce an annual sustainability report.[13] And Anadarko Petroleum shareholders asked that company to assess the impact of its business on climate change.[14]

Shareholder resolutions need not pass—and often need not even come to a vote—to steer a corporation's practices in a new direction. The adverse publicity of a pointed shareholder resolution may be enough to pressure corporate executives to change course. The Executive Director of the Interfaith Center on Corporate Responsibility (ICCR) in New York notes that the stream of CEOs into her office has increased over three decades as ICCR has become more effective at gaining support for shareholder resolutions. She now spends roughly half of her time meeting with corporate executives who are responding to proposed resolutions.[15]

The smallest dimension of SRI is community investing, which involves steering investment capital to areas that traditionally lack it, such as inner cities in wealthy countries or microfinance cooperatives in developing countries. Community investing moves beyond screened investments, which aim to green existing economic activity, and focuses instead on generating entirely new nodes of sustainable economic activity. In the United States, community investing has grown from $4 billion in 1995 to $20 billion in 2005.[16]

SRI performance has often been competitive with standard investments. A 2005 study of the Domini 400 Index—a measure of the performance of SRI portfolios—found that returns over the long run were very competitive with those of the S&P 500.[17] Meanwhile, the top 10 eco-fund performers globally in 2006, with investments in eight countries, posted average returns

Table 1. Socially Responsible Investments, by Region, Mid-2000s

Country or Region	Socially Responsible Investments	Year of Data
	(billion dollars)	
United States	2,290.0	2005
Europe	1,224.0	2005
Canada	439.0	2006
Australia/New Zealand	7.0	2005
Japan	2.6	2007

Source: Social Investment Forum, Eurosif, Social Investment Organization, Corporatemonitor, and Cangen Biotechnologies.

of 39 percent.[18]

Indeed, SRI—once synonymous with inferior returns—is becoming a respectable investment source as evidence mounts that sustainable business practices often help a company's bottom line (and, in turn, make the firm an attractive target for green investments). Research indicates that sustainable practices increase a firm's value in concrete ways: by cutting waste and therefore costs, helping to recruit and retain the best staff, strengthening revenues, and reducing liability risk associated with unsustainable practices (such as emitting carbon, a major contributor to climate change).[19] These advantages may help explain why the portfolio value of the "Global 100" most sustainable companies, unveiled each year at the World Economic Forum in Davos, Switzerland, outperformed the MSCI World index (a common benchmark) by 80 percent over the period January 2000 to December 2005.[20]

Pension funds are increasingly important in raising SRI investment totals. In the United Kingdom in 2000, for example, occupational pension schemes were required to reveal whether they took social, environmental, or ethical factors into account when deciding what stocks to invest in.[21] Similar regulations have since been passed in Australia, Sweden, and Germany.[22] Similarly, the California Public Employees Retirement System, one of the largest pension funds in the United States, com-

mitted in 2001 to sell its tobacco stocks, to screen investments to ensure they meet human rights, labor, and environmental standards, and to dedicate some 2 percent of its assets to community investment.[23] And the nearly eightfold growth in SRI value in Canada between 2004 and 2006 was largely due to a shift of public-sector pension funds toward SRI investments.[24] Pension funds are often the largest group of institutional shareholders and carry considerable weight in determining how companies act.[25]

Retail investment firms are increasingly helping clients make socially responsible investments. Beyond offering screened investment options, some retailers are promoting particular segments of SRI, such as microfinance, which provides very small loans to impoverished, entrepreneurially minded individuals, primarily in developing countries. In 2006, TIAA-CREF—a U.S. firm with $380 billion in assets under management—established a $100-million fund for microcredit that offers investors a way to direct their investments to capital-short sectors.[26]

SRI is also spurred by growing recognition at the highest levels internationally. In 2006 the United Nations launched the Principles of Responsible Investment, which commits signatories to apply environmental, social, and governance norms to their investment practices.[27] The launch featured more than 70 institutional investors as charter signatories, representing more than $4.5 trillion in assets.[28] The principles grew out of work promoting SRI in other U.N. programs, including the U.N. Environment Programme Finance Initiative and the U.N. Global Compact.[29] In addition, the International Interfaith Investment Group launched in 2005 is a global effort to steer institutional religious wealth toward sustainable projects.[30] The group includes 16 organizations from four world faiths as members.[31]

Health Features

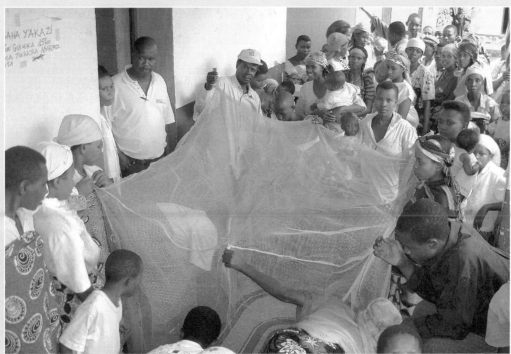

© 2006 Isabelle Walhin, Courtesy of Photoshare

Community health workers demonstrate how to set up an anti-malarial mosquito bed net, Burundi.

▶ HIV/AIDS Continues Worldwide Climb

▶ Malaria Remains a Threat

▶ Male Reproductive Health Declines

For data and analysis on these health topics and on others such as infant mortality, cigarette production, and obesity, go to www.worldwatch.org/vsonline.

HIV/AIDS Continues Worldwide Climb

Lindsay Hower Jordan

As of December 2006, some 39.5 million people around the world were living with HIV, the virus that causes AIDS—37.2 million of them were adults, with an estimated 17.7 million women over the age of 15 carrying the infection.[1] Some 4.3 million people were newly infected with HIV in 2006, with a little more than 500,000 of those new infections occurring in people under the age of 15.[2] In 2006, 2.9 million people died from AIDS.[3]

In Western Europe, HIV infection rates increased sharply in the last eight years, from 42 cases per million people in 1998 to 74 cases per million in 2006.[4] Around the Baltic region, the high HIV infection trend that characterized the turn of the twenty-first century there appears to be abating, particularly in Latvia and Estonia.[5] South and Southeast Asia is home to 7.8 million infected individuals, an 8-percent increase from 7.2 million in 2004; in East Asia, the figure is around 750,000 people, up from 620,000 in 2004; and in Latin America, the infected total is around 1.7 million, up from 1.5 million in 2004.[6]

Sub-Saharan Africa is home to nearly two thirds of people worldwide living with HIV.[7] (See Figure 1.) In this region, there were 2.8 million newly infected individuals in 2006, up slightly from 2.6 million just two years earlier.[8] At the end of 2006, UNAIDS estimated that 24.7 million sub-Saharan Africans are infected with HIV, an increase of 1.1 million since 2004.[9] Seventy-two percent of deaths due to AIDS occur in sub-Saharan Africa.[10]

Modes of HIV transmission vary widely by region. For example, injecting drug users account for 67 percent of all HIV cases in Eastern Europe and Central Asia.[11] In South and Southeast Asia, in contrast, they account for 22 percent of cases, while 49 percent of victims there are infected through commercial sex work (8 percent are sex workers; 41 percent are clients).[12] Men having sex with men accounted for 4 percent of HIV cases in Eastern Europe and Central Asia, 5 percent in South and Southeast Asia, and 26 percent in Latin America.[13]

In 2006 there were more women infected

LINKS pp. 50, 108

with HIV in every region of the world than ever before.[14] Women are at particularly high risk in countries with rampant infection rates, since they are not traditionally in a position of power or decisionmaking in their sexual relationships. In the Caribbean, North Africa, Oceania, and the Middle East, almost half the adults infected with HIV are women age 15 or older.[15]

In sub-Saharan Africa, women outnumber men in infection estimates, accounting for up to 60 percent of people living with HIV.[16] According to Ludfine Anyango, national HIV/AIDS coordinator at Action Kenya-International, "many women cannot even choose when to have sex or not. Many cannot ask their husbands to use a condom because in addition to being thought as unfaithful, they fear being beaten. The woman then has no choice but to continue having unprotected sex with her spouse."[17] Street violence likewise exposes female sex workers to high risk of HIV infection for the same reasons, according to Ros Sokunthy of Women's Agenda for Change, a Cambodia-based organization fighting to protect women's rights, including those of female sex workers.[18]

In 70 countries surveyed, use of testing and counseling services has quadrupled since 2001, from 4 million to 16.5 million people in 2005.[19] In Sudan, where HIV prevalence in North Africa is at its highest, 350,000 people—1.6 percent of the country's population—were living with HIV in 2005.[20] Current knowledge of the benefits of contraception and of how HIV is transmitted is pitifully poor there: in a 2005 survey of police officers in Khartoum, only 2 percent of the men knew that condoms could prevent transmission.[21] Certain countries, including Iran, have implemented clean syringe and methadone operations as well as government-funded clinics that offer free HIV counseling, testing, and treatment.[22]

From 1996 to 2005, funding for HIV/AIDS assistance efforts in low- and middle-income countries increased from $300 million to $8.3 billion.[23] But current trends in existing pledges may indicate the funding is waning, with pledges totaling just $8.9 billion in 2006 and

$10 billion in 2007.[24] The United Nations has projected needs at $14.9 billion in 2006, $18.1 billion in 2007, and $22.1 billion in 2008, highlighting a sustained gap between current funds and future needs over the next few years.[25] In August 2006, the Bill & Melinda Gates Foundation committed $500 million over five years to the Global Fund to Fight AIDS, Tuberculosis and Malaria—the largest gift to support AIDS and other disease research from a nongovernmental source since the fund was established.[26]

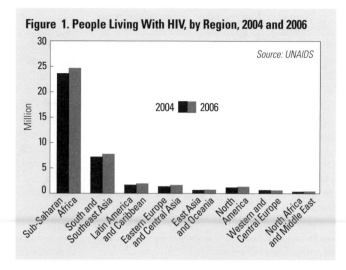

Figure 1. People Living With HIV, by Region, 2004 and 2006

Source: UNAIDS

In 2000, after settling a lawsuit by the South African government on patent rights, the leading producers of HIV medicines established the Accelerating Access Initiative (AAI) in collaboration with five U.N. agencies, including UNAIDS, to provide more anti-retroviral medicines at lower costs.[27] An AAI report in June 2003 indicated that the number of Africans receiving treatment under this initiative was eight times higher than when the program began in 2000, totaling approximately 75,000.[28] By March 2005, AAI was reaching more than 427,000 patients.[29]

Pharmaceutical companies have made considerable strides in working with corporate firms and national governments to craft national efforts that address HIV infection and alleviate stress from limited access to drugs.[30] Public-private partnerships are an encouraging development, such as Johnson & Johnson's royalty-free collaboration with the International Partnership for Microbicides—a cross-sector partnership that aims to develop and distribute its recently developed compound TMC120 as an experimental vaginal microbicide.[31]

Generic anti-retroviral drugs are beginning to dominate the global consumer drug market. In 2006, preliminary statistics suggested that 70 percent of anti-retrovirals in Nigeria, Haiti, and Zambia were generic.[32] This influx is primarily

the result of the U.S. Food and Drug Administration's approval of 29 generic AIDS drugs.[33]

In the 1990s, Brazil had an HIV rate that rivaled South Africa's, but since 1996 it has cut the infection rate to 0.6 percent of the adult population—including an 80-percent reduction in HIV-related hospitalization—by becoming the first country to offer universal treatment.[34] But as HIV patients build resistance to old drugs and as drug companies refuse to offer contracts for newer generic versions, Brazil was forced to spend 75 percent more on anti-retrovirals between 2004 and 2006.[35] Although the government has negotiated with drug companies for the cheapest price outside of Africa, it still has to pay $17,000 a year per patient—a jarring price tag for a government accustomed to buying older generic drugs for hundreds of dollars per patient annually.[36]

Malaria Remains a Threat

Mary Galinski and Esmeralda Meyer

The World Health Organization (WHO) reports that 350–500 million people get malaria annually, with at least 1 million of these cases resulting in death.[1] This is astounding for a disease that by and large is preventable and treatable.[2] Worse yet, new research indicates that these numbers may be an underestimate.[3]

Malaria—which dates back to ancient times—is endemic in 107 countries and territories today, the result of a vicious cycle of transmission of the *Plasmodium* parasite from female anopheline mosquitoes to humans and back to the mosquito.[4] Breaking this cycle is the key to controlling and eliminating the disease.[5] Its symptoms include intense fever, sweats, chills, headache, and nausea.[6] In serious cases, severe anemia, organ failure, and coma are possible, with death a major concern in the absence of effective drug treatment and clinical care. Children, pregnant woman, and non-immune adults are the most vulnerable individuals.[7]

The direct costs for countries with the highest incidence of malaria has been estimated at $1.9 billion annually, while the global figure is $3.2 billion.[8] Lost work and school days and a high level of morbidity affect individuals, families, and communities worldwide on a grand scale.[9] A recent World Economic Forum report on malaria drives home these points and concludes that the private sector can contribute significantly to malaria control by investing in local programs.[10]

LINKS p. 108

Today there is an unprecedented move to scale up interdisciplinary approaches, coordination, and the use of multiple proven malaria control tools in sub-Saharan Africa, where 46 countries suffer from some of the highest levels of malaria.[11] To aid the process in Africa and elsewhere, the new Malaria Atlas Project is developing a global database on the prevalence of this disease, taking into account the geographical distribution patterns and transmission characteristics of the two predominant species, *P. falciparum* and *P. vivax*.[12] (See Figure 1.)

In 1897, Sir Ronald Ross discovered that malaria was transmitted by mosquitoes.[13] From 1915 to 1952, the Rockefeller Foundation developed projects to control malaria, and from 1955 to 1965 WHO led a Global Malaria Eradication Campaign.[14] Both organizations focused on eliminating the mosquito vector, with notable successes in the reduction of malaria.[15] But the campaign, based on DDT spraying, was abandoned soon after the publication of Rachel Carson's *Silent Spring* in 1962, which led many to believe that DDT should not be used—regardless of its public health benefits.[16]

Now the usefulness of DDT has again been recognized, following two years of advocacy that began in 1999 with an open letter by the Malaria Foundation International that had 416 signatories from 63 countries.[17] In December 2000, the importance of DDT for malaria control was recognized in the Stockholm Convention on Persistent Organic Pollutants.[18] In 2006, WHO gave DDT a clean bill of health, and financial backing for DDT spraying resumed.[19]

At the time of World War II, chloroquine became recognized as a cheap and effective "wonder drug" to cure malaria, at 10¢ per treatment.[20] Yet chloroquine and all subsequent malaria medications have developed resistance or reduced sensitivity, especially for treating the most lethal form of malaria, caused by *P. falciparum*.[21] Combination drugs are now recommended to stop the spread of resistance.[22]

In 2004, a report in the medical journal *The Lancet* claimed that it was medical malpractice to use malaria drugs that had a high chance of being ineffective.[23] In turn, the Global Fund to Fight AIDS, Tuberculosis and Malaria—the major funding source today for malaria drugs—vowed to support only the use of reliable antimalarial drugs and to seek more than $1 billion from donors to pay for artemisinin-based combination therapy (ACT).[24] The Fund has determined that $2.9 billion is needed in 2007 to use all available means to control malaria, yet only about $300 million is currently allocated.[25] Today, experts are working to produce more artemisinin, to develop adequate amounts of effective ACT, to manage and reduce the high cost of this drug, and to develop avenues for its effective distribution and use.[26] An intermittent therapy for pregnant women is also recognized as a priority.[27] With the higher cost of current malaria treatments, the confirmation of malaria diagnosis by microscopy or rapid diagnostic tests is crucial for control and prevention strategies.[28]

Long-lasting insecticide-treated bed nets are now being promoted as a way to prevent malaria through the distribution of millions of nets in Africa and as a tool to gain the attention of the public and raise new funds.[29] Sleeping with the protection of these nets will help prevent the disease.[30] While it is not a total solution, it is a reasonable line of attack in light of today's interdisciplinary approach to combating malaria.[31]

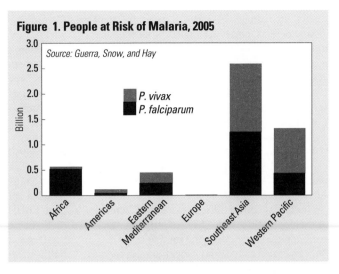

Figure 1. People at Risk of Malaria, 2005

Source: Guerra, Snow, and Hay

- P. vivax
- P. falciparum

Fifteen years ago malaria received little if any media attention worldwide. Modern approaches to malaria advocacy and education began with the 1995 launch of the Malaria Foundation International's Web site.[32] Dozens of organizations and initiatives are now rallying around this cause.[33] While funding is far from adequate and malaria is still not covered well in the media, attention has been increasing steadily.[34] Control measures are being implemented in Africa with increasingly large funds from the Bill & Melinda Gates Foundation (over $765 million since 1999 for malaria research and control), the World Bank ($500 million promised in 2000), and the President's Malaria Initiative ($1.2 billion pledged by the U.S. government in 2005).[35]

Some successes in malaria control have been noted. The Roll Back Malaria program in Eritrea evaluated the use of insecticide-treated bed nets, DDT spraying, and case management between 2000 and 2004 and reported an 84-percent decline in malaria morbidity and a 40-percent drop in case fatality.[36] In another study, researchers found that high coverage with ACT was the most cost-effective strategy for malaria control in sub-Saharan Africa.[37]

To encourage a greater commitment, in 2005 the Malaria R&D Alliance—a global coalition of research and development organizations—surveyed malaria research and development investments.[38] It found that only $323 million was dedicated to malaria research in 2004, less than 0.3 percent of total health research spending worldwide.[39]

More than 90 percent of health research resources are spent on diseases that affect just 10 percent of the world's population, while research is badly needed on new malaria drugs, on development of malaria vaccines, and on ways to use and monitor malaria control tools effectively.[40] Many potential malaria vaccines are being developed, although none have reached the market to date.[41] But knowledge about malaria genome sequences, modern technologies, and concerted efforts provide hope that an effective vaccine will be available in the future.[42]

Male Reproductive Health Declines

Peter Stair

Rising rates of testicular cancer, more frequent genital defects, and deteriorating sperm quality indicate that male reproductive health is declining in many populations. Between 1960 and 2002, closely monitored men in Europe, the United States, and New Zealand became two to seven times more likely to suffer from testicular cancer (see Table 1) and, on average, produced about half as many sperm per milliliter of semen.[1] These changes have yet to be adequately explained, but their rapid onset among younger men suggests environmental disruptions during early development are at least partly to blame.

Testicular cancer afflicts less than 1 percent of the population, but it has become the most common malignancy among men ages 20 to 34.[2] The rate of testicular cancer among men under 50 across northern Europe, Australia, New Zealand, and the United States has been increasing about 2–4 percent a year since the 1960s.[3] Since rates of cryptorchidism (undescended testicles) and hypospadias (shortened urinary tracts) have risen simultaneously, some specialists have identified a broader "testicular dysgenesis syndrome" that threatens male fertility.[4] Men born with testicular deformities and survivors of testicular cancer tend to have more problems producing enough healthy sperm to conceive children.[5]

LINKS p. 50

The most recent analysis of several dozen studies conducted primarily in Europe, North America, and Australia since the 1930s found that sperm density has fallen from 110–170 sperm per milliliter to just under 60.[6] Studies demonstrating such a broadly based decline in sperm counts have been controversial, however, because sperm quality can vary widely over the course of a man's life—rising during periods of abstinence, for example, and declining during the summer.[7] Study samples have also often come from men seeking vasectomies (who tend to have higher than average sperm counts) or from men in couples experiencing infertility (who tend to have lower than average).[8]

Yet there is a consensus that sperm counts vary by region and have fallen more in some places than others. Men in New York City have sperm counts 75 percent higher than men in Columbia, Missouri, for example, while men in Turku, Finland, have counts 25 percent higher than men in Copenhagen, Denmark.[9] Although just one sperm is required to fertilize an egg, researchers have identified sperm concentrations of 40 per milliliter of semen as the threshold below which men's fertility declines.[10] According to a 2006 analysis, about 40 percent of men from Denmark and Norway are below this level.[11]

In the United States, testicular cancer is most common among urbanites in the northwestern and central states and less likely in New England and the South.[12] In Europe, the incidence is greatest in the region encompassing Denmark and Switzerland and lowest in the Baltic states, France, Italy, and Spain.[13] In some cases this inter-regional difference is sharp: men born in Denmark or Norway are three times as likely to have testicular cancer as men born in Estonia or Finland.[14]

Some populations may be genetically more vulnerable to reproductive disruption than others. Testicular cancer is five times less common among African Americans, for example, while European men, on average, have higher sperm counts than American or Japanese men.[15] But such differences do not explain the degree of geographic variation in reproductive maladies: although they are closely related in genetic backgrounds, men in Denmark have notably lower sperm counts that men in southern Sweden.[16]

Some lifestyle choices affect sperm counts. Men who drink more alcohol or smoke more cigarettes tend to have lower sperm counts.[17] Overweightness, age, and other drug use have also been associated with lower sperm quality.[18] Cell phones may inhibit healthy sperm production: a 2007 study found that men in Cleveland, New Orleans, and Mumbai who used cell phones longer than four hours each day had sperm counts 25 percent lower than those who never used them.[19] Yet this relationship may be confounded by other variables, such as sedentary living.[20]

Exposure to chemicals that interfere with sex hormones remains the prevailing explanation for the increase in male reproductive disor-

Table 1. Testicular Cancer Rates in Selected Countries or States, 1960–2002

Country or Region	1960	1970	1980	1990	1995	2002
			(cases per 100,000 men aged 20–34)			
State of New York	2.1	2.3	4.1		5.0	
State of Iowa		3.7	4.1	4.5	4.9	
United States						5.5
Denmark	3.8	4.9	7.8	9.2	9.9	10.3
Finland	1.2	1.1	1.5	2.5	2.7	3.2
Netherlands	2.8		3.2	4.0	4.7	5.8
Norway	3.3	4.4	5.9	8.0	8.2	10.6
Slovenia	1.3	1.9	2.8	4.3	5.8	8.6
Sweden	2.4	2.5	3.3	4.8	5.0	5.8
New Zealand	3.5	3.7	5.3	5.6	5.8	6.4

Source: World Health Organization.

ders.[21] Scientists have identified more than 50 synthetic chemicals that disrupt the endocrine system and more than a dozen additional suspects.[22] Those most firmly associated with reproductive disorders include dioxins, which are released during paper pulp processing, coal combustion, and waste incineration; polychlorinated biphenyls, which are used for a range of electrical, insulation, lubrication, and other industrial purposes; and pesticides that are commonly used in agriculture.[23]

Phthalates—a common plastic softener—have also been linked with reproductive maladies.[24] A 2006 study in China found that workers exposed to phthalates while manufacturing polyvinyl chloride materials had lower levels of testosterone.[25] In a broader Massachusetts study, men with higher levels of phthalate metabolites in their blood had lower sperm counts, lower sperm motility, and more sperm deformities.[26]

Many compounds known to be disruptive to reproductive development have been banned—but only after years of widespread use. Between 1950 and 1975, for example, doctors prescribed the estrogen-mimic diethylstilbestrol to 5 million pregnant women, hoping to promote fetal growth and prevent spontaneous abortions.[27] Two decades passed before researchers realized the sons of these women were more likely to have smaller testicles, genital deformities, and impaired sperm quality.[28]

Today more than 80,000 synthetic chemicals are in production, and most have unknown long-term effects.[29] Acknowledging this, in 1996 the U.S. Environmental Protection Agency initiated an Endocrine Disruption Screening Program to evaluate more than 15,000 chemicals.[30] In Europe, similar concerns culminated in the 2005 Prague Declaration on Endocrine Disruption, which was signed by hundreds of scientists from Europe and North America. It warned of "serious risks" to men's fertility and urged more comprehensive monitoring of male reproductive maladies.[31]

Clouding researchers' ability to identify harmful chemicals is the potential for some chemicals to be safe in isolation but dangerous in tandem with others. A 2006 study of tadpoles found that only 4 percent died when they were exposed to each of nine common pesticides alone but 35 percent died from exposure to a mixture of all nine.[32] Since each person on Earth now contains detectable levels of several hundred synthetic chemicals, in varying proportions, it is impossible to identify all the potentially toxic chemical cocktails. Accordingly, the Prague Declaration called for a "precautionary approach" to regulating potentially disruptive chemicals—an appeal to err on the side of caution even in the absence of scientific consensus about the sources of endocrine disruption.[33]

Notes

GRAIN PRODUCTION FALLS AND PRICES SURGE (pages 20–21)

1. U.N. Food and Agriculture Organization (FAO), *FAOSTAT Statistical Database*, at faostat.fao.org, updated 23 January 2007; FAO, *Food Outlook* (Rome: December 2006), pp. 50, 51, 55.
2. FAO, *FAOSTAT*, op. cit. note 1.
3. Per capita figures calculated from ibid. and from U.S. Bureau of the Census, *International Data Base*, electronic database, Suitland, MD, updated 24 August 2006.
4. FAO, *FAOSTAT*, op. cit. note 1; Census Bureau, op. cit. note 3.
5. FAO, *Food Outlook*, op. cit. note 1, p. 2.
6. Ibid., pp. 4, 5.
7. Ibid.
8. U.S. Department of Agriculture (USDA), Foreign Agricultural Service, "Tightening World Grain Supplies Push Prices to Decade High Levels," in *Grain: World Markets and Trade* (Washington, DC: November 2006); FAO, *Food Outlook*, op. cit. note 1, p. 11.
9. FAO, *Food Outlook*, op. cit. note 1, p. 13.
10. Ibid.
11. Ibid.
12. Ibid.; FAO, *FAOSTAT*, op. cit. note 1.
13. FAO, *Food Outlook*, op. cit. note 1, p. 9; U.S. share from ibid., p. 52.
14. Dr. Keith Collins, chief economist, USDA, "Advancing Renewable Energy: An American Rural Renaissance," U.S. Agriculture and the Emerging Bioeconomy, presentation, 12 October 2006.
15. Renewable Fuels Association, "October Ethanol Production Ties All-Time High Yearly Production, Demand Up More than 25 Percent," press release (Washington, DC: 29 December 2006).
16. USDA, op. cit. note 8.
17. FAO, *Food Outlook*, op. cit. note 1, p. 4.
18. Ibid., p. 8.
19. Ibid., p. 10.
20. FAO, "Cereal Prices Surge to Highest Levels in Decade, Strong Implications for Meat and Other Agricultural Commodities," press release (Rome: 7 December 2006).
21. Ibid.

SOYBEAN DEMAND CONTINUES TO DRIVE PRODUCTION (pages 22–23)

1. U.N. Food and Agriculture Organization (FAO), *FAOSTAT Statistical Database*, at faostat.fao.org, updated 6 February 2007.
2. Ibid.
3. Ibid.
4. Ibid.
5. Ibid
6. U.S. Department of Agriculture (USDA), "Soybean Backgrounder," at www.ers.usda.gov/publications/OCS/apr06/OCS200601_Lowres.pdf, April 2006.
7. Ibid.
8. Ibid.
9. FAO, op. cit. note 1.
10. Ibid.
11. Ibid.
12. USDA, Production Estimates and Crop Assessment Division, Foreign Agricultural Service, "Brazil: 2005/06 Soybean Area Projected to Decline," at www.fas.usda.gov/pecad/highlights/2005/09/brazil_12sep2005/, 12 September 2005.
13. Ibid.
14. FAO, op. cit. note 1.
15. World Wide Fund for Nature (WWF), "Problems: Forest Conversion," at www.panda.org/about_wwf/what_we_do/ forests/problems/conversion/index.cfm, 18 June 2006.
16. KeShun Liu, *Soybeans: Chemistry, Technology, and Utilization* (Gaithersburg, MD: Aspen, 1999), p. 25.
17. USDA, op. cit. note 6.
18. Liu, op. cit. note 16.
19. USDA, op. cit. note 6.

20. Ibid.; FAO, op. cit. note 1.
21. USDA, op. cit. note 6; FAO, op. cit. note 1.
22. USDA, op. cit. note 6.
23. Jorge Fernandez-Cornejo and Margriet Caswel, *The First Decade of Genetically Engineered Crops in the United States* (Washington, DC: Economic Research Service, USDA, 2006).
24. Ibid.; Oliver Batch, "Seeds of Dispute," (London) *The Guardian*, 22 February 2006; FAO, op. cit. note 1.
25. USDA, op. cit. note 6.
26. FAO, *Food Outlook* (Rome: December 2006), p. 20; WWF, op. cit. note 15.
27. USDA, op. cit. note 6.
28. "Brazil Biodiesel Production Set to Reach 1.3 Bln Litres by July," *F.O. Licht's World Ethanol & Biofuels Report*, 22 February 2007.

MEAT OUTPUT AND CONSUMPTION GROW (pages 24–25)

1. Food and Agriculture Organization (FAO), "Meat and Meat Products," *Food Outlook* (Rome: December 2006); Figure 1 from ibid. and from FAO, "Meat and Meat Products," *Food Outlook* (Rome: June 2006).
2. Henning Steinfeld and Pius Chilonda, "Old Players, New Players," *Livestock Report 2006* (Rome: FAO, 2006), p. 3.
3. FAO, June 2006, op. cit. note 1.
4. FAO, December 2006, op. cit. note 1.
5. Ibid.
6. FAO, June 2006, op. cit. note 1; FAO, December 2006, op. cit. note 1.
7. Ibid.
8. Ibid.
9. FAO, December 2006, op. cit. note 1.
10. Ibid.
11. Ibid.
12. FAO, *Livestock Report 2006* (Rome: 2006); Henning Steinfeld et al., *Livestock's Long Shadow, Environmental Issues and Options* (Rome: FAO, 2006).
13. FAO, *FAOSTAT Statistical Database*, at faostat.fao.org, updated 24 January 2006; Compassion in World Farming, *Laying Hens Fact Sheet*, revised January 2004, at www.ciwf.org.uk/publications/Factsheets/Factsheet%20-%20Laying%20Hens%20.pdf.
14. Michael Greger, *Bird Flu: A Virus of Our Own Hatching* (New York: Lantern Books, 2006), p. 111–12, 113.

15. Steinfeld et al., op. cit. note 12, pp. xx–xxi.
16. Ibid., p. 4.
17. Ibid., p. xxi.
18. Ibid.
19. Ibid.
20. Ibid., p. xxii.
21. Ibid., p. xx.
22. FAO, *Pollution from Industrialized Livestock Production*, Policy Brief 2 (Rome: Livestock Information, Sector Analysis, and Policy Branch, Animal Production and Health Division, undated).
23. Ibid.

SEAFOOD INCREASINGLY POPULAR AND SCARCE (pages 26–27)

1. U.N. Food and Agriculture Organization (FAO), *FAOSTAT Statistical Database*, at faostat.fao.org, updated 15 September 2006. The United Nations recently revised the way it totals seafood to include seaweeds and aquatic plants. So although the total seems to be 15 million tons higher than reported in *Vital Signs 2006–2007*, there has not actually been a jump in total fish production.
2. FAO, op. cit. note 1.
3. Ibid.
4. B. Worm et al., "Impacts of Biodiversity Loss on Ocean Ecosystem Services," *Science*, 3 November 2006, pp. 787–90.
5. FAO, op. cit. note 1; U.S. Bureau of the Census, *International Data Base*, electronic database, Suitland, MD, updated 24 August 2006.
6. FAO, op. cit. note 1; Census Bureau, op. cit. note 5.
7. FAO, op. cit. note 1.
8. FAO, op. cit. note 1; Census Bureau, op. cit. note 5.
9. FAO, op. cit. note 1; Census Bureau, op. cit. note 5.
10. FAO, op. cit. note 1; Census Bureau, op. cit. note 5.
11. Nanna Roos et al., "Fish and Health," in Corinna Hawkes and Marie T. Ruel, eds., *Understanding the Links Between Agriculture and Health for Food, Agriculture, and the Environment*, 2020 Focus No. 13 (Washington, DC: International Food Policy Research Institute (IFPRI), May 2006).
12. FAO, op. cit. note 1.
13. Figure of 30 percent from Meryl Williams, "The Transition in the Contribution of Living Aquatic Resources to Food Security," Food, Agriculture, and the Environment Discussion Paper 13 (Washington,

DC: IFPRI, 1996); 6 percent from FAO, op. cit. note 1.

14. FAO, op. cit. note 1.

15. Farmed versus wild fish from FAO, op. cit. note 1; feed use from Jackie Alder and Daniel Pauly, *On the Multiple Uses of Forage Fish: From Ecosystems to Markets*, Fisheries Centre Research Reports (Vancouver, BC: Sea Around Us Project, Fisheries Centre, University of British Columbia, 2006).

16. Alder and Pauly, op. cit. note 15.

17. FAO, *State of World Aquaculture 2006* (Rome: 2006).

18. FAO, op. cit. note 1.

19. Ibid.

20. Alder and Pauly, op. cit. note 15, pp. vii, 3.

21. Ibid.

22. Employment from FAO, *The State of World Fisheries and Aquaculture* (Rome: 2004), pp. 22, 24.

23. Ibid.

24. Catch per fisher from FAO, *The State of World Fisheries and Aquaculture* (Rome: 1998); more exacting fishing techniques from Daniel Pauly and Jay Maclean, *In a Perfect Ocean* (Washington, DC: Island Press, 2006), p. 72.

IRRIGATED AREA STAYS STABLE (pages 28–29)

1. U.N. Food and Agriculture Organization (FAO), Land Use (1961–2003) Data Collections, *FAOSTAT Statistical Database*, at faostat.fao.org, viewed 15 February 2007.

2. Ibid.

3. Ibid.; U.S. Bureau of the Census, *International Data Base*, electronic database, Suitland, MD, updated 24 August 2006.

4. FAO, *Unlocking the Water Potential of Agriculture* (Rome: 2003).

5. FAO, op. cit. note 1.

6. Ibid.

7. J. M. Faurèsa et al., *The FAO Irrigated Area Forecast for 2030*, Research Paper (Rome: FAO, 2000).

8. Songcai Yong, "Irrigated Area Mapping for China," presented to International Workshop on Global Irrigated Area Mapping, Colombo, Sri Lanka, 25–27 September 2006.

9. FAO, op. cit. note 4.

10. Peter H. Gleick, "Water for Food: How Much Will be Needed?" in Peter H. Gleick, *The World's Water 2000–2001* (Washington, DC: Island Press, 2000), p. 80.

11. Sandra Postel, "Redesigning Irrigated Agriculture," in Worldwatch Institute, *State of the World 2000* (New York: W. W. Norton & Company, 2000), p. 40.

12. FAO, op. cit. note 4.

13. Postel, op. cit. note 11, p. 41.

14. FAO, op. cit. note 4.

15. Ibid.

16. Ibid.; FAO, *The State of Food and Agriculture in Asia and the Pacific* (Bangkok: 2006), p. 13.

17. Lester R. Brown, *Eco-Economy* (New York: W. W. Norton & Company, 2001), p. 154.

18. Yong, op. cit. note 8; 64 percent from China's National Bureau of Statistics, *China's Environment Statistics 2005* (Beijing: 2005).

19. FAO, op. cit. note 4.

20. Sandra Postel, *Pillar of Sand* (New York: W. W. Norton & Company, 1999), p. 87.

21. Ibid.

22. UNDP, "Poor Farmers Face Double Water Crisis: Climate Change and Competition," at hdr.undp.org/hdr2006/press/releases/english/RP4-HDR06_PR4E.pdf, February 2007.

23. Postel, op. cit. note 11, p. 48.

24. FAO, op. cit. note 16, p. 14.

25. FAO, op. cit. note 4.

26. Postel, op. cit. note 11, p. 48.

27. FAO, op. cit. note 4.

28. Lester R. Brown, *Plan B 2.0* (New York: W. W. Norton & Company, 2006), p. 167.

FOSSIL FUEL USE UP AGAIN (pages 32–33)

1. International Energy Agency (IEA), *Oil Market Report*, 13 March 2007, p. 5; near-record in 2004 from BP, *Statistical Review of World Energy* (London: 2005), p. 2. (The increase in 2004 was the largest since 1976).

2. Worldwatch estimate based on data through 2003 from BP, *Statistical Review of World Energy* (London: 2006), p. 12, and on annual growth rates for 2004 through 2006 derived from IEA, op. cit. note 1, p. 5.

3. Based on data from IEA, op. cit. note 1, p. 44.

4. Ibid.

5. "Expensive Gas Equals Less Driving," *CNN.com*, 2 December 2006; "US Oil Demand Dropped Slightly in 2006," *MSNBC.com*, 19 January 2007.

6. IEA, op. cit. note 1, p. 6.

Notes

7. Fourth consecutive year from U.S. Department of Energy (DOE), Energy Information Administration (EIA), *Short Term Energy Outlook*, 7 March 2007, from DOE, EIA, *Short Term Energy Outlook*, 7 February 2006, and from Platts, cited in BP, op. cit. note 1, p. 14; production declines from IEA, *Oil Market Report*, 13 February 2007; political developments from Steven Mufson, "Factors That Pushed Oil Price Up Are Now Pushing It Down," *Washington Post*, 16 September 2006.

8. DOE, 7 March 2007, op. cit. note 7; DOE, 7 February 2006, op. cit. note 7. Data reflect West Texas intermediate prices and are presented in constant dollars.

9. Barbara Hagenbaugh, "Global Tension Pushes Oil to Record," *USA Today*, 13 July 2006.

10. IEA, op. cit. note 1, p. 45.

11. "Huge Oil Find in Gulf of Mexico," *BBC News Online*, 6 September 2006; Brazil and Middle East from the American Association of Petroleum Geologists, "Important Discoveries Opened Horizons–'06 Charts Some Remarkable Finds," *Explorer*, January 2007.

12. IEA, op. cit. note 1, p. 45.

13. Ibid.

14. Ibid.; Christine A. Telyan and Konstantin Kovalenko, *Russian Crude Oil Production Outlook for 2006: Is a Further Slowdown in Store?* (Cambridge, MA: Cambridge Energy Research Associates, March 2006).

15. Iran production from IEA, op. cit. note 1; analyst projections from Carola Hoyos, "National Oil Company of Iran," *Financial Times*, 25 March 2007.

16. Based on data from BP, op. cit. note 2, p. 28.

17. Ibid.

18. Ibid.

19. Ibid., p. 35.

20. Ibid.

21. Ibid.

22. Erik Shuster and Scott Klara, National Energy Technology Laboratory, DOE, "Tracking New Coal-Fired Power Plants: Coal's Resurgence in Electric Power Generation," 24 January 2007, PowerPoint slide 4.

23. Steven James, "US Coal-Fired Power Plant Plans up in Smoke?" *Reuters*, 4 March 2007; "Midwest Has 'Coal Rush,' Seeing No Alternative—Energy Demand Causes Boom in Plant Construction," *Washington Post*, 10 March 2007.

24. BP, op. cit. note 2, p. 35.

25. "India's Coal Demand May Quadruple by 2031— Minister," *Reuters*, 13 March 2007.

26. BP, op. cit. note 2, p. 41.

27. Winnie Zhu, "China Meets Less Than a Third of Energy Saving Target," *Bloomberg.com*, 28 February 2007.

28. "China's Interest in the Middle East," *The Energy Bulletin*, 2 November 2006.

29. Wu Zhong, "Auto Boom Worsens China's Energy Crunch," *Asia Times*, 3 June 2006.

30. National Bureau of Statistics of China, *The Statistics Report on China's National Economy and Social Development in 2006* (Beijing: February 2007); "Vehicle Market Forecast: China's Auto Consumption Enters a New Stage of Private Purchase," *Xinhua News Agency*, 28 February 2007.

31. "China's Power Generating Capacity Tops 622 Mln Kilowatts," *People's Daily Online*, 24 January 2007; Peter Fairley, "China's Coal Future," *Technology Review*, 4 January 2007.

32. IEA, *World Energy Outlook 2006* (Paris: 2006), Summary and Conclusions, pp. 1–5.

33. Ibid.

34. See, for example, Greenpeace and European Renewable Energy Council, *Energy [R]evolution: A Sustainable World Energy Outlook* (Amsterdam and Brussels: January 2007); World Business Council for Sustainable Development, *Pathways to 2050 Energy and Climate Change* (Geneva: 2005); German Advisory Council on Global Change, *World in Transition: Towards Sustainable Energy Systems* (London: Earthscan, 2004); Royal Dutch/Shell, *The Evolution of the World's Energy Systems* (London: 1996).

35. "EU Agrees Renewable Energy Target," *BBC News Online*, 9 March 2007.

NUCLEAR POWER VIRTUALLY UNCHANGED (pages 34–35)

1. Installed nuclear energy capacity is defined as reactors connected to the grid as of 31 December 2006 and is based on Worldwatch Institute database compiled from statistics from the International Atomic Energy Agency; from press reports, primarily from *Associated Press*, *Reuters*, and *World Nuclear Association (WNA) News Briefing*; and from Web sites.

2. Worldwatch Institute database, op. cit. note 1; see other Vital Signs in this volume for growth rates of

other sources.

3. Worldwatch Institute database, op. cit. note 1.

4. Ibid.

5. Ibid.

6. Ibid.

7. Ibid.

8. Ibid.

9. U.S. Department of Energy, Energy Information Administration, *Commercial Nuclear Power 1990: Prospects for the United States and the World* (Washington, DC: September 1990), pp. 101–06.

10. "Tennessee Valley Authority to Purse 2 New Reactors," *Washington Post*, 29 January 2007.

11. Christopher Kirkpatrick, "Duke Energy CEO Reveals Doubts about Nuclear Plant," *Charlotte Observer*, 20 January 2007.

12. Ken Silverstein, "Nuclear Energy's Potential Comeback," *Energy Biz Insider*, 22 January 2007.

13. "S&P Expects More Nuclear Plants, But Not Before 2015," *Platts*, 26 June 2006.

14. "TVO's Olkiluoto-2's Commercial Operation Delayed to 2010, 2011," *Platts*, 5 December 2006.

15. Satu Hassi, European Parliament member, e-mail to Christopher Flavin, 19 February 2007.

16. "France: Electricite de France Has Begun Awarding Contracts," *WNA News Briefing*, 5 September 2006.

17. Michael Settle, "Blair Warned Against Rushing to Build Nuclear Power Plants MPs Accuse Ministers," *The Herald*, 10 July 2006.

18. "Nuclear Power in China" (December 2006), "Nuclear Power in India" (November 2006), and "Nuclear Power in Korea" (November 2006), World Nuclear Association, at www.world-nuclear.org/info/inf81.html, viewed 27 January 2007.

19. Worldwatch Institute database, op. cit. note 1.

20. Ibid.

21. "North Korea: The Korean Peninsula Energy Development Organization," *WNA News Briefing*, 6 June 2006; Bo-Mi Lim, "North Korea Asked to Pay US$1.9 Billion for Aborted Nuclear Power Plant Project," *Associated Press*, 16 January 2007.

22. United Nations, Security Council Resolution 1737, 23 December 2006.

23. "Russia to Launch Iranian Nuclear Power Plant on Schedule," *Associated Press*, 28 January 2007.

24. "USA: Congress Approved the Bill for the Civilian Nuclear Deal with India," *WNA News Briefing*, 12 December 2006.

WIND POWER STILL SOARING (pages 36–37)

1. Global Wind Energy Council (GWEC), "Global Wind Energy Markets Continue to Boom—2006 Another Record Year," press release (Brussels: 2 February 2007); Birger Madsen, BTM Consult, e-mail to author, 8 March 2007.

2. Installations in 2006 from GWEC, op. cit. note 1, and from "Global Wind Energy Market Grew 32 Pct in 2006—VDMA," *Reuters*, 5 February 2007; 2005 installations calculated by Worldwatch with data from GWEC, op. cit. note 1, and from GWEC, "Record Year for Wind Energy: Global Wind Power Market Increased by 43% in 2005," press release (Brussels: 17 February 2005).

3. European Wind Energy Association (EWEA), "European Market for Wind Turbines Grows 23% in 2006," press release (Brussels: 1 February 2007); United States from GWEC, op. cit. note 1.

4. Climate change concerns from Intergovernmental Panel on Climate Change, *Climate Change 2007: The Physical Science Basis. Summary for Policymakers* (Geneva: 2007); energy security concerns from Steve Hargreaves, "Wind Power Blows Through China," CNNMoney.com, 20 February 2007.

5. Additions in 2006 from GWEC, op. cit. note 1; Figure 3 data from ibid., from GWEC, op. cit. note 2, from BTM Consult ApS, from EWEA, from American Wind Energy Association (AWEA), from Bundesverband WindEnergie (BWE), from Paul Gipe of AWEA, and from Instituto para la Diversificacion y Ahorro Energetico.

6. Total capacity from "U.S. Wind Power to Grow 26 Percent in 2007—AWEA," *Reuters*, 24 January 2007.

7. Stephen Lacey, "AWEA: Stalled Midwest Wind Projects a Minor Setback," RenewableEnergyAccess.com, 27 October 2006; EWEA, "Texas Leads Another Busy Year," *Wind Directions*, September 2006, p. 6.

8. AWEA, "Wind Power Capacity in U.S. Increased 27% in 2006 and Is Expected to Grow an Additional 26% in 2007," press release (Washington, DC: 23 January 2007); AWEA, "Legislative Affairs," available at www.awea.org/legislative/#RPS.

9. "U.S. Wind Power to Grow 26 Percent in 2007," op. cit. note 6.

10. GWEC, op. cit. note 1; EWEA, op. cit. note 3.

11. Installations and total capacity from EWEA, op. cit.

Notes

note 3; share of European Union demand from GWEC, op. cit. note 1.

12. GWEC, op. cit. note 1; GWEC, "German Wind Energy Market Still Leads the Global Market," *Newsletter*, 15 February 2007.

13. GWEC, op. cit. note 1; EWEA, op. cit. note 3.

14. "German Renewable Energy Usage at Record High in 2006," *Reuters*, 5 January 2007.

15. GWEC, op. cit. note 1; GWEC, "Spain Reaches 11,615 MW of Installed Wind Power," *Newsletter*, 16 February 2007.

16. GWEC, op. cit. note 1.

17. Ibid.; change from 2005 calculated by Worldwatch with data from GWEC, op. cit. note 1, and from EWEA, "Wind Power Installed in Europe by End of 2005 (Cumulative)," from www.ewea.org; not on track from GWEC, op. cit. note 15.

18. GWEC, op. cit. note 1.

19. Over 50 from GWEC and Greenpeace, *Global Wind Energy Outlook 2006* (Brussels: September 2006), p. 5; 13 countries from GWEC, op. cit. note 1.

20. GWEC, op. cit. note 1.

21. Ibid.

22. Ibid.

23. Ibid.

24. Leader in small wind from Yao Runping, "China to Become World's Largest Wind Power Producer," *Xinhua*, 16 January 2007; sixth place from GWEC, op. cit. note 1.

25. GWEC, op. cit. note 1.

26. Redouble by 2010 from GWEC, "China Set to More Than Treble Wind Power Capacity by 2010," *Latest News*, 18 January 2007; 2020 from Yao, op. cit. note 24.

27. Ron Pernick, "Clean Energy Markets: Managing High-Tech Growth," *CE Views*, at www.cleanedge.com.

28. Larry Flowers, U.S. National Renewable Energy Laboratory, e-mail to Christopher Flavin, Worldwatch Institute, 26 March 2007; Christine Real de Azua, AWEA, e-mail to author, 27 March 2007.

29. John Donnelly, "Two Oil Giants Plunge into the Wind Business: Shell, BP Intend to Play Major Role," *Boston Globe*, 2 March 2007.

30. Figure of $20 billion from "Global Wind Energy Market," op. cit. note 2, and from GWEC, op. cit. note 1; 2016 from Joel Makower, Ron Pernick, and Clint Wilder, *Clean Energy Trends 2007* (Oakland, CA, and Portland, OR: March 2007), p. 3.

31. GWEC and Renewable Energy Systems Ltd., "Wind Power Ready to Meet Looming Energy Gap, Says Report," press release (Brussels: 5 September 2006).

SOLAR POWER SHINING BRIGHT (pages 38–39)

1. *PV News* (Prometheus Institute), April 2007, p. 8.

2. Global growth rate based on data from Travis Bradford, Prometheus Institute, discussions with author, 2 April, 6 April, and 10 April 2007; global total from Travis Bradford, Prometheus Institute, e-mails to author, 5 April, 7 April, and 8 April 2007.

3. Based on *PV News*, op. cit. note 1, p. 8, and on Paul Maycock, *PV News*, various issues.

4. *PV News*, op. cit. note 1, p. 8.

5. Based on data from *PV News*, op. cit. note 1, p. 8; China from ibid., p. 9, and from *Marketbuzz 2007*, cited in "Solarbuzz Reports World Solar Photovoltaic Market Growth of 19% in 2006," Solarbuzz.com, 19 March 2007.

6. Increase over 2005 based on *Marketbuzz 2006*, cited in "Solarbuzz Reports World Solar Photovoltaic Market Grew 34% in 2005," Solarbuzz.com, 15 March 2006; 2006 installations from Bradford, discussions, op. cit. note 2.

7. Germany from Bradford, e-mails, op. cit. note 2, and from Paul Gipe, "Germany Installed 1,150 MW of Photovoltaics in 2006 Says Magazine," *Renewable EnergyAccess.com*, 12 March 2007; more than half based on Bradford, e-mails, op. cit. note 2.

8. Installations in 2005 from Paul Gipe, "Feed Law Powers Germany to New Renewable Energy Record," *RenewableEnergyAccess.com*, 5 February 2007.

9. *PV News*, op. cit. note 1, pp. 8–9.

10. Growth and rankings based on ibid.

11. Production and share of global market based on *PV News*, op. cit. note 1, pp. 8–9; half of the expansion from *PV News* (Prometheus Institute), March 2007, p. 1.

12. "World Cell Production Grows 40% in 2006," *PV News* (Prometheus Institute), March 2007, p. 6.

13. Exports from Bradford, discussions, op. cit. note 2; domestic installations from Bradford, e-mails, op. cit. note 2.

14. Based on *PV News*, op. cit. note 1, p. 8.

15. *PV News*, op. cit. note 11, p. 6; also www.firstsolar .com.

16. *PV News*, op. cit. note 11, p. 6.
17. Ranking and capacity installed based on Bradford, e-mails, op. cit. note 2; increase over 2005 from "US Grid-Tied Installations Grow 60%," *PV News* (Prometheus Institute), March 2007, p. 3.
18. "US Grid-Tied Installations," op. cit. note 17.
19. Bradford, discussions, op. cit. note 2.
20. Environmental Technologies Action Plan, "Spain's New Building Energy Standards Place the Country Among the Leaders in Solar Energy in Europe," September 2006, at ec.europa.eu/environment/etap; "Spain Makes Solar Panels a Must on New Buildings," *Reuters*, 14 November 2006.
21. *PV News*, op. cit. note 11, pp. 1, 6.
22. *PV News*, op. cit. note 1, pp. 1, 2; "Silicon Valley Goes Solar," *Associated Press*, 25 September 2006.
23. Michael Liebreich, New Energy Finance, cited in "2007 Starts Strongly for Clean Energy Investment," press release (London: New Energy Finance Ltd., 2 April 2007).
24. Prometheus Institute forecast, *PV News*, op. cit. note 1, p. 2.
25. Ibid.; Bradford, discussions, op. cit. note 2.
26. "Honda Enters Solar Cell Market with New Subsidiary," RenewableEnergyAccess.com, 6 December 2006.
27. Bradford, discussions, op. cit. note 2; Bradford, e-mails, op. cit. note 2.
28. Bradford, e-mails, op. cit. note 2.
29. Ibid.

BIOFUEL FLOWS SURGE (pages 40–41)

1. All fuel ethanol data and the 2002–06 biodiesel figures from Christoph Berg, senior analyst, F.O. Licht, e-mails to Rodrigo G. Pinto, 20–22 March 2007; 2000–01 biodiesel numbers from "World Biodiesel Production to Cross 4 Mln Tonne Mark in 2006," *F.O. Licht's World Ethanol & Biofuels Report*, 25 April 2006; 1991–99 biodiesel figures from Christoph Berg, senior analyst, F.O. Licht, e-mail to Peter Stair, Worldwatch Institute, 25 January 2006. F.O. Licht has revised its worldwide fuel ethanol production figures for 2005 and prior years due to the fact that the largest part of Brazilian ethanol exports did not end up in the fuel market but in beverage and industrial applications; the downward adjustment was substantial in relation to a previous preliminary

estimate for 2005. For comparison purposes, there are 159 liters per barrel. There are 1,262 liters of ethanol or 1,136 liters of biodiesel in a ton. While a liter of ethanol contains approximately 68 percent as much energy as a liter of gasoline, a liter of biodiesel contains roughly 87 percent as much energy as a liter of diesel.
2. Berg, e-mails 20–22 March 2007, op. cit. note 1.
3. Ibid.
4. Biofuel production numbers from ibid.; oil production numbers from International Energy Agency (IEA), *Oil Market Report*, 13 March 2007, page 45.
5. Berg, e-mails 20–22 March 2007, op. cit. note 1; "World Ethanol Production Unfazed by Recent Oil Price Weakness," *F.O. Licht's World Ethanol & Biofuels Report*, 2 November 2006.
6. Berg, e-mails 20–22 March 2007, op. cit. note 1; "Fuel Ethanol Balance–December 2006," *F.O. Licht's World Ethanol & Biofuels Report*, 7 March 2007.
7. "World of Biofuels 2006," *F.O. Licht's World Ethanol & Biofuels Report*, 10 January 2007; "Brazil News: Dec 2006 Ethanol Exports Up By 71% Y/Y–Trade Ministry," *F.O. Licht's World Ethanol & Biofuels Report*, 10 January 2007; "Biofuels and World Agricultural Trade," *F.O. Licht's World Ethanol & Biofuels Report*, 9 August 2006; "World Biofuel Markets Moving in Tandem at Seville," *F.O. Licht's World Ethanol & Biofuels Report*, 26 May 2006; "Fuel Ethanol Balance – December 2006," op. cit. note 6; Berg, e-mails 20–22 March 2007, op. cit. note 1. Data disaggregating Brazil's fuel ethanol exports from ethanol exports generally were not available by the time of publication.
8. Berg, e-mails 20–22 March 2007, op. cit. note 1.
9. Renewable Fuels Association, "U.S. Fuel Ethanol Industry Biorefineries and Production Capacity," Washington, DC, 3 April 2007; C. F. Runge and B. Senauer, "How Biofuels Could Starve the Poor," *Foreign Affairs*, in press.
10. Production data from Berg, e-mails 20–22 March 2007, op. cit. note 1; production capacity data from "Biofuels Output to Rise Further–Commission," *F.O. Licht's World Ethanol & Biofuels Report*, 8 October 2006; "World of Biofuels 2006," op. cit. note 7; "Bright Prospects for Biodiesel in South America," *F.O. Licht's World Ethanol & Biofuels Report*, 23 October 2006; "Asia—A Major Vegetable Oil Market Gears Up For Biodiesel," *F.O. Licht's World Ethanol &*

Notes

Biofuels Report, 29 January 2007; "World Biodiesel Production to Cross 4 Mln Tonne Mark," op. cit. note 1. Shares and rankings calculated based on these figures.

11. Production data from Berg, e-mails 20–22 March 2007, op. cit. note 1.

12. Suzanne Hunt and Peter Stair, "Biofuels Hit a Gusher," in Worldwatch Institute, *Vital Signs 2006–2007* (New York: W. W. Norton & Company, 2006), p. 40.

13. "World of Biofuels 2006," op. cit. note 7; Justin Gillis, "New Fuel Source Grows on the Prairie," *Washington Post*, 22 June 2006; "Myanmar Leapfrogs to Oil Independence Through Biofuels Program—Questions about Human Rights Remain," *Biopact*, 13 August 2006; Monte Reel, "U.S. Seeks Partnership With Brazil on Ethanol: Countering Oil-Rich Venezuela Is Part of Aim," *Washington Post*, 8 February 2007; "Castro Hits Out At US Ethanol Use," *BBC News*, 29 March 2007.

14. "Investors Now Running with Biofuels Ball," *F.O. Licht's World Ethanol & Biofuels Report*, 8 June 2006; "US Venture Capital Boom for Biofuels Sector," *F.O. Licht's World Ethanol & Biofuels Report*, 7 March 2007.

15. U.S. stocks figure from "Biofuels and World Agricultural Trade," op. cit. note 7; venture capital figures from "US Venture Capital Boom for Biofuels Sector," op. cit. note 14.

16. Gillis, op. cit. note 13; Gerard Wynn, "Shell Plans Cleaner Second Generation of Biofuels," *Reuters*, 7 October 2006; Alexei Barrionuevo, "6 Get Grants from U.S. to Support Bio-Refineries," *New York Times*, 1 March 2007; Mariana Durão, "Energias Alternativas Para Celulose," *Jornal do Commercio (RJ)*, 2 April 2007. Cellulosic materials include stems, leaves, and wood.

17. Preliminary estimate from "Brasil Entrega a Energia da Biomassa a Multinacionais," *Exame*, 7 July 2006.

18. "Ethanol Boom More than Just a Pipedream," *F.O. Licht's World Ethanol & Biofuels Report*, 21 November 2006; "World of Biofuels 2006," op. cit. note 7; "World Biofuel Markets Moving in Tandem," op. cit. note 7; Gerard Wynn and Muriel Boselli, "Success Derails Biofuels Bandwagon," *Reuters*, 6 March 2007.

19. "FAPRI Projections Plot Declining Margins in Bioenergy," *F.O. Licht's World Ethanol & Biofuels Report*, 7 March 2007; Patrick Barta, "Hot Biofuel Stocks May

Cool; Analysts Say Demand, Environment Worries Could Play Factor," *Wall Street Journal*, 20 November 2006; "World of Biofuels 2006," op. cit. note 7; Lauren Etter, Ilan Brat, and Steven Gray, "Corn's Rally Sends Ripples; Ethanol Boom Cheers Grain Farmers, Pinches Food Makers," *Wall Street Journal*, 18 January 2007; "Technological Advances Support Growth," *F.O. Licht's World Ethanol & Biofuels Report*, 6 February 2007; "World Biodiesel Production Growth May Slow in 2007," *F.O. Licht's World Ethanol & Biofuels Report*, 22 March 2007.

20. "Hot Biofuel Stocks May Cool," op. cit. note 19; "World of Biofuels 2006," op. cit. note 7; Patrick Barta and Jane Spencer, "Crude Awakening: As Alternative Energy Heats Up, Environmental Concerns Grow; Crop of Renewable 'Biofuels' Could Have Drawbacks; Fires Across Indonesia; Palm-Oil Boom Ignites Debate," *Wall Street Journal*, 5 December 2006; "Energy Companies Rethink Palm Oil as Biofuel," *International Herald Tribune*, 27 March 2007; Runge and Senauer, op. cit. note 9.

21. See M. Verdonk, C. Dieperink, and A. P. C. Faaij, "Governance of the Emerging Bio-Energy Markets," *Energy Policy*, in press; Jinke van Dam et al., "Overview of Recent Developments in Sustainable Biomass Certification," IEA Bioenergy Task 40 Paper (Paris: IEA, 22 December 2006).

22. "The German Biofuel Quota Law? A Model for the European Union?" *F.O. Licht's World Ethanol & Biofuels Report*, 14 December 2006; Sustainable Production of Biomass Project Group, *Criteria for Sustainable Biomass Production*, Final Report (Sittard, Netherlands: Energy Transition Task Force, Government of Netherlands, July 2006).

23. "Investors Now Running with Biofuels Ball," op. cit. note 14; "Hot Biofuel Stocks May Cool," op. cit. note 19; "Recuo do Petróleo Ameaça Investimento em Biocombustíveis," *Folha de São Paulo*, 22 January 2007; "China: Construction of New Corn-Based Ethanol Plant Suspended," *F.O. Licht's World Ethanol & Biofuels Report*, 10 January 2007; "Biodiesel Faz Cotação do Sebo Dobrar," *Gazeta Mercantil*, 12 February 2007; Scott Kilman, "DuPont-BP Venture Will Make Competing Product to Ethanol," *Wall Street Journal*, 21 June 2006; "Virgin's Newest Venture Targets Alternative Fuel Development," *Jet Fuel Intelligence*, 11 December 2006; "NASA and Boeing Join Brazil to Develop Biokerosene Aviation Fuel," *Bio-*

pact, 30 August 2006; Carey Gillam, "Dung Power at U.S. Ethanol Plant," *Reuters*, 26 February 2007. Dedicated sources of biofuel include both crops and waste streams not widely used for other ends. Conceptually between fuel ethanol and gasoline, the promising biobutanol contains as much energy per liter as gasoline.

24. "Ethanol Boom More than Just a Pipedream," op. cit. note 18; "Myanmar Leapfrogs," op. cit. note 13.

25. "Ethanol Boom More than Just a Pipedream," op. cit. note 18.

26. Alexandra Olson, "U.S., Brazil Launch Biofuels Forum," *Washington Post*, 3 March 2007.

27. "Global Bioenergy Partnership Secretariat Up, Running," *FAO Newsroom*, 25 September 2006.

28. "A Closer Look at Africa's 'Green OPEC'," *Biopact*, 2 August 2006.

CARBON EMISSIONS CONTINUE UNRELENTING RISE (pages 42–43)

1. Steven Piper, Scripps Institution of Oceanography, University of California, San Diego, e-mail to Una Song, Worldwatch Institute, 13 March 2007.

2. Increase since 1959 from C. D. Keeling et al., "Atmospheric CO_2 Concentrations (ppmv) Derived from In Situ Air Samples Collected at Mauna Lao Observatory, Hawaii," Scripps Institution of Oceanography, University of California, San Diego, May 2005, and from Piper, op. cit. note 1; pre-industrial value was 280 parts per million, per Intergovernmental Panel on Climate Change (IPCC), *Climate Change 2007: The Physical Science Basis—Summary for Policymakers* (Geneva: February 2007), p. 2.

3. IPCC, op. cit. note 2, p. 2.

4. Calculated by Worldwatch with data from BP, *Statistical Review of World Energy* (London: 2006), and from G. Marland et al., "Global, Regional, and National Fossil Fuel CO_2 Emissions," in Carbon Dioxide Information Analysis Center, *A Compendium of Data on Global Change* (Oak Ridge, TN: Oak Ridge National Laboratory, U.S. Department of Energy, 2006).

5. Calculated by Worldwatch with data from BP, op. cit. note 4, and from Marland et al., op. cit. note 4.

6. U.S. share calculated by Worldwatch: global total estimated with data from Marland et al., op. cit.

note 4, and from BP, op. cit. note 4; U.S. emissions from U.S. Energy Information Administration, Office of Integrated Analysis and Forecasting, *Emissions of Greenhouse Gases in the United States 2005* (Washington, DC: November 2006), p. 13.

7. Lauren Morello, "Climate: U.S. Electricity Demand Spurred 0.8 Percent Rise in '04 Emissions—EPA," *Environment and Energy News*, 20 February 2007.

8. Alister Doyle, "Greenhouse Gases Hit New High, May Be Asia Growth," *Reuters*, 19 February 2007.

9. Worldwatch calculation, based on BP, op. cit. note 4; surpassing United States before 2010 according to International Energy Agency, *World Energy Outlook 2006* (Paris: 2006), Executive Summary, p. 5.

10. IPCC, op. cit. note 2.

11. Number of experts from Michael Byrnes, "Interview—Scientist Says Sea Level Rise Could Accelerate," *Reuters*, 13 March 2007; IPCC, op. cit. note 2, pp. 8, 10.

12. Alister Doyle, "Tropical Losers, Northern Winner for Warming?" *Scientific American*, 3 April 2007; Andrew C. Revkin, "Poor Nations to Bear Brunt as World Warms," *New York Times*, 1 April 2007.

13. Land-ocean index from J. Hansen et al., "Global Land-Ocean Temperature Index in .01C, Base Period 1951–1980 (January-December)," Goddard Institute for Space Studies, at www.giss.nasa.gov/data/update/gistemp/GLB.Ts+dSST.txt.

14. Australia from Andrew C. Revkin, "Global Warming Trend Continues in 2006, Climate Agencies Say," *New York Times,* 15 December 2006; China from "Shanghai Has Warmest Winter on Record," *Reuters*, 2 March 2007; United Kingdom from Ian Sample, "This Year Will Be Britain's Warmest Since Records Began, Say Scientists," *Guardian*, 14 December 2006.

15. U.S. National Oceanic and Atmospheric Administration (NOAA), "NOAA Says U.S. Winter Temperature Near Average, Global December-February Temperature Warmest on Record," press release (Washington, DC: 15 March 2007).

16. Hansen et al., op. cit. note 13.

17. Peter N. Spotts, "New Search for Global Warming at Poles," *Christian Science Monitor*, 26 February 2007; Lauren Morello, "Climate: Int'l Research Effort 'to Unlock Mysteries of the Poles'," *Greenwire*, 26 February 2007.

18. IPCC, op. cit. note 2, p. 8.

19. Deborah Zabarenko, "Global Warming Is a Human

Notes

Rights Issue—Nobel Nominee," *Reuters*, 5 March 2007.

20. September ice from IPCC, op. cit. note 2, p. 8, and from National Climatic Data Center, NOAA, "Climate of 2006—Annual Report, 11 January 2007, at www.ncdc.noaa.gov/oa/climate/research/2006/ann/global.html.

21. "Climate: This Winter Was World's Warmest on Record, Feds Say," *Greenwire*, 16 March 2007; "The Senate's Task on Warming" (editorial), *New York Times*, 6 January 2007.

22. "Senate's Task on Warming," op. cit. note 21.

23. Nicholas Stern, *The Economics of Climate Change: The Stern Review* (Cambridge, UK: Cambridge University Press, 2006), Summary of Conclusions and Executive Summary.

24. Michelle Nichols, "Climate Change as Dangerous as War—UN Chief Ban," *Reuters*, 2 March 2007.

25. Darren Samuelsohn, "Climate: Lawmakers Begin Looking Abroad for Clues on Emission Curbs," *Greenwire*, 26 March 2007.

26. International Emissions Trading Association and World Bank, *State and Trends of the Carbon Market 2006, Update: (January 1-September 30, 2006)* (Washington, DC: October 2006), Executive Summary.

27. First nine months of 2006 from International Emissions Trading Association and World Bank, op. cit. note 26; 2005 total from Heather Timmons, "Data Leaks Shake Up Carbon Trade," *New York Times*, 16 May 2006; China and India from The Climate Institute, "Join the $30 Billion International Carbon Market," press release (Sydney, Australia: 7 February 2007).

28. "Climate: EU Commits to Emissions, Renewable Energy Targets," *Greenwire*, 9 March 2007.

29. States from David Ammons, "Lawmakers Join Al Gore's Crusade to Ease Climate Change," *Seattle Post-Intelligencer*, 10 March 2007; Timothy Gardner, "Investors to Press US Congress on Global Warming," *Reuters*, 19 March 2007.

WEATHER-RELATED DISASTERS CLIMB
(pages 44–45)

1. Munich Reinsurance Company (Munich Re), Nat-CatSERVICE, e-mail to author, 16 March 2007. Worldwatch calculation based on Center for Research on the Epidemiology of Disasters (CRED), *EM-DAT: The OFDA/CRED International Disaster Database*, viewed 27 March 2007. CRED data are continuously revised; the calculations include extreme temperature events, floods, slides, wildfires, and windstorms. This calculation corrects for multiple country entries per disaster. Munich Re and CRED use different criteria and methodologies to calculate disaster incidence.

2. Munich Re, op. cit. note 1.

3. Munich Re, "Munich Re's New 'Topics Geo — Natural Catastrophes 2006' Published," press release (Munich: 6 March 2007).

4. "2005 Atlantic Hurricane Season," and "2006 Atlantic Hurricane Season," National Hurricane Center, at www.nhc.noaa.gov.

5. Worldwatch calculation based on CRED, op. cit. note 1; Munich Re reported 13,600 weather-related disaster deaths in 2006 and acknowledges that the toll in some regions may be underrepresented.

6. Worldwatch calculation based on CRED, op. cit. note 1.

7. "UN Food Agency Provides Assistance to Almost 120,000 Affected by Flooding in Ethiopia," UN News Centre, 22 August 2006.

8. International Federation of Red Cross and Red Crescent Societies (IFRC), "China: Red Cross Responds to Worst Typhoon in Half a Century," press release (Geneva: 11 August 2006).

9. Worldwatch calculation based on CRED, op. cit. note 1; affected refers here to all those injured, made homeless, or otherwise affected by a disaster.

10. Worldwatch calculation based on CRED, op. cit. note 1.

11. IFRC, *World Disasters Report 2006* (Geneva: 2006), p. 27.

12. Richard Kerr, "Is Katrina a Harbinger of Still More Powerful Hurricanes?" *Science*, 16 September 2005, p. 1807.

13. Farid Dahdouh-Guebas et al., "How Effective Were Mangroves as a Defence Against the Recent Tsunami?" *Current Biology*, vol. 15, no. 12 (2005), pp. 443–47.

14. John Young, "Black Water Rising," *World Watch*, September/October 2006, p. 30.

15. Reinhard Mechler and Joanne Linnerooth-Bayer with David Peppiatt, *Disaster Insurance for the Poor?* (Geneva: ProVention Consortium/International Institute for Applied Systems Analysis, 2006), p. 6.

16. Ibid.
17. Sachs quoted in Mica Rosenberg, "Economist Pushes Insurance as Answer to Disasters," *Reuters AlertNet*, 11 April 2006.
18. Rosenberg, op. cit. note 17.
19. Ibid.
20. United Nations Office for the Coordination of Humanitarian Affairs, "The Central Emergency Response Fund," at ochaonline.un.org/webpage.asp?Page=2101.
21. United Nations Office for the Coordination of Humanitarian Affairs, "CERF Pledges and Contributions," at ochaonline2.un.org/Default.aspx?tabid=8693, updated 27 March 2007.

OZONE LAYER STABILIZING BUT NOT RECOVERED (pages 46–47)

1. Craig Long, U.S. National Oceanic and Atmospheric Administration (NOAA), e-mails to author, 29 November 2006, 3 March 2007, and 20 March 2007.
2. Ibid.; World Meteorological Organization (WMO) and U.N. Environment Programme (UNEP), *Scientific Assessment of Ozone Depletion: 2006* (Geneva: 2006), p. 3.1. (The downward trend from 1979 to 1996 reflects impacts due to solar variation and aerosols ejected from volcanoes El Chicon (1984) and Pinatubo (1991). Solar variation from 1996 to the present may account for the slight decline in 2003–06).
3. Long, op. cit. note 1.
4. WMO and UNEP, "New Report Projects Later Recovery of Ozone Layer But Early Signs of Mending Confirm Effectiveness of Montreal Protocol on Substances That Deplete the Ozone Layer," press release (Geneva: 18 August 2006).
5. Stephen Leahy, "Ailments Surge as Ozone Hole Widens," *Inter Press Service*, 11 November 2006.
6. Ibid.
7. UNEP, *The GEO Data Portal*, at geodata.grid.unep.ch.
8. Ibid.
9. David W. Fahey, "Twenty Questions and Answers About the Ozone Layer: 2006 Update," part of WMO and UNEP, op. cit. note 2, Q.1, Q.5.
10. WMO and UNEP, op. cit. note 2, pp. 1.1–1.2.
11. Ibid., p. 7.1.
12. Rita Beamish, "U.S. Pesticide Stockpile Under Scrutiny at World Ozone Meeting in India," *Environmental News Network*, 3 November 2006.
13. "Whistle Blown on Illegal CFC Trade," *BBC News*, 31 January 2004; "The Price of Keeping Cool In Asia," *New York Times*, 23 February 2007; RAL Quality Assurance, "Residual CFCs Are a Ticking 'Time Bomb'," at PeopleandPlanet.net, 28 September 2006.
14. WMO and UNEP, op. cit. note 2, p. 6.3.
15. Ibid.
16. Long, op. cit. note 1.
17. Fahey, op. cit. note 9, p. Q.30.
18. Ibid.
19. Long, op. cit. note 1.
20. Ibid.

POPULATION RISE SLOWS BUT CONTINUES (pages 50–51)

1. U.S. Bureau of the Census, *International Data Base*, electronic database, Suitland, MD, updated 24 August 2006; U.N. Population Division, *World Population Prospects: The 2004 Revision, Volume III* (New York: United Nations, 2006).
2. Census Bureau, op. cit. note 1; U.N. Population Division, op. cit. note 1.
3. Census Bureau, op. cit. note 1; U.N. Population Division, op. cit. note 1.
4. U.N. Population Division, op. cit. note 1.
5. Ibid.
6. Ibid.
7. Ibid.
8. Population Reference Bureau (PRB), *World Population Data Sheet 2006* (Washington, DC: 2006).
9. Ibid.; U.N. Population Division, op. cit. note 1.
10. U.N. Population Division, op. cit. note 1.
11. Ibid.
12. Ibid.
13. PRB, op. cit. note 8.
14. U.N. Population Division, op. cit. note 1.
15. Ibid.
16. World Wildlife Fund (WWF), *Living Planet Report 2006* (Gland, Switzerland: WWF, Zoological Society of London, and Global Footprint Network, 2006), p. 28.
17. Ibid.
18. U.N. Population Division, op. cit. note 1.
19. Molly Moore, "As Europe Grows Grayer, France

Notes

Devises Baby Boom," *Washington Post*, 18 October 2006.

20. Ibid.

21. U.N. Population Division, *World Population Prospects: The 2004 Revision* (New York: 2005).

22. David Satterthwaite and Gordon McGranaham, "Providing Clean Water and Sanitation," in Worldwatch Institute, *State of the World 2007* (New York: W. W. Norton & Company, 2007), p. 27.

23. Ibid.

24. United Nations Population Fund (UNFPA), *State of World Population 2005* (New York: 2005).

25. World Health Organization (WHO), UNICEF, and UNFPA, *Maternal Mortality in 2000: Estimates Developed by WHO, UNICEF, and UNFPA* (Geneva: WHO, 2003).

WORLD IS SOON HALF URBAN (pages 52–53)

1. U.N. Population Division, *World Urbanization Prospects 2005* (New York: 2006), also available online at esa.un.org/unup. This Vital Sign is based on Kai N. Lee, "An Urbanizing World," in Worldwatch Institute, *State of the World 2007* (New York: W. W. Norton & Company, 2007), pp. 3–21. Molly O'Meara Sheehan also contributed to this research.

2. Figure 1 from ibid.

3. Ibid.

4. Ibid.

5. Ibid.

6. Ibid.

7. Ibid.

8. Megacities from ibid.; National Research Council (NRC), *Cities Transformed: Demographic Change and Its Implications in the Developing World* (Washington, DC: National Academies Press, 2003), pp. 95–99.

9. Africa and Figure 2 from U.N. Population Division, op. cit. note 1.

10. NRC, op. cit. note 8, pp. 99–102.

11. Ibid.

12. UN-HABITAT, *State of the World's Cities 2006/7* (London: Earthscan, 2006), p. 16.

13. U.N. Population Division, op. cit. note 1.

14. NRC, op. cit. note 8, pp. 102–06; polluted cities in China from World Bank, cited in "A Great Wall of Waste—China's Environment," *The Economist*, 21 August 2004.

15. India's urban poverty from UN-HABITAT, op. cit.

note 12, p. 11.

16. U.N. Population Division, op. cit. note 1, viewed August 2006.

17. Figure 3 and share of total from ibid.

18. Ibid.

19. NRC, op. cit. note 8, p. 107.

20. UN-HABITAT, op. cit. note 12.

21. For analysis of this trend, see Gordon McGranahan et al., *The Citizens at Risk: From Urban Sanitation to Sustainable Cities* (Sterling, VA: Earthscan, for Stockholm Environment Institute, 2001), chapter 4.

22. Kirk R. Smith and Majid Ezzati, "How Environmental Health Risks Change with Development: The Epidemiologic and Environmental Risk Transitions Revisited," *Annual Review of Environment and Resources*, November 2005, pp. 291–333.

23. Xuemei Bai and Hidefumi Imura, "A Comparative Study of Urban Environment in East Asia: Stage Model of Urban Environmental Evolution," *International Review for Environmental Strategies*, summer 2000, pp. 135–58; McGranahan et al., op. cit. note 21.

24. Millennium Ecosystem Assessment, "Summary for Decision-Makers," in *Ecosystems and Human Well-Being: Synthesis* (Washington, DC: Island Press, 2005), p. 1; McGranahan et al., op. cit. note 21.

25. Herbert Girardet, *Cities People Planet* (Chichester, U.K.: John Wiley & Sons, 2004), pp. 123–25; Herbert Girardet, *The Gaia Atlas of Cities* (London: Gaia Books, 1992), pp. 22–23.

26. Ken Yeang, *Bioclimatic Skyscrapers* (London: Ellipsis London Press, 2000).

ECONOMY AND STRAIN ON ENVIRONMENT BOTH GROW (pages 54–55)

1. International Monetary Fund (IMF), *World Economic Outlook Database* (Washington, DC: September 2006). Note the 2006 figure is a preliminary estimate from September 2005 and is subject to change. These figures represent inflation-adjusted IMF data.

2. IMF, op. cit. note 1.

3. Ibid. Note: unless otherwise specified, all further analysis is based on PPP terms.

4. IMF, op. cit. note 1.

5. Ibid.; IMF, *World Economic Outlook 2006: Financial Systems and Economic Cycles* (Washington, DC: Sep-

tember 2006), p.49.

6. Christopher Flavin and Gary Gardner, "China, India, and the New World Order," in Worldwatch Institute, *State of the World 2006* (New York: W. W. Norton & Company, 2006), p. 7.

7. Jonathan Watts, "Major Pollution Spill 'Every Other Day' in China," (London) *The Guardian*, 11 January 2007; Jianqiang Liu, "The 'Special Interests' Destroying China's Environment," *China Dialogue*, 24 January 2007.

8. IMF, op. cit. note 1. Note: because of economic and geographic similarities between the former Soviet states and Mongolia, the IMF includes Mongolia in its analysis of these economies.

9. IMF, op. cit. note 5, pp. 62–73.

10. IMF, op. cit. note 1.

11. IMF, op. cit. note 5, p. 41.

12. Ibid.

13. IMF, op. cit. note 1.

14. Ibid.; IMF, op. cit. note 5, pp. 44–45; "Eurozone 2006 Economic Growth Trimmed to 2.6 pct," *EU Business*, 6 March 2007.

15. IMF, op. cit. note 5, pp. 44–45.

16. IMF, op. cit. note 1; IMF, op. cit. note 5, pp. 47–48.

17. IMF, op. cit. note 1; U.S. Bureau of the Census, *International Data Base*, electronic database, Suitland, MD, updated 24 August 2006.

18. IMF, op. cit. note 1; Census Bureau, op. cit. note 17.

19. IMF, op. cit. note 1; Census Bureau, op. cit. note 17.

20. John Talberth, Clifford Cobb, and Noah Slattery, *The Genuine Progress Indicator 2006, A Tool for Sustainable Development* (Oakland, CA: Redefining Progress, 2006).

21. Ibid.

22. Millennium Ecosystem Assessment, *Ecosystems and Human Well-being: Synthesis* (Washington, DC: Island Press, 2005), p. 1.

23. Global Footprint Network, *National Footprint and Biocapacity Accounts*, 2006 edition (Oakland, CA: 2006); World Wide Fund for Nature (WWF), Zoological Society of London, and Global Footprint Network, *Living Planet Report 2006* (Gland, Switzerland: WWF, 2006).

24. Global Footprint Network, op. cit. note 23.

25. Calculation based on Global Footprint Network, op. cit. note 23, and on WWF, Zoological Society of London, and Global Footprint Network, op. cit. note 23.

STEEL PRODUCTION SOARS (pages 56–57)

1. International Iron and Steel Institute (IISI), "World Produces 1,239.5 mmt of Crude Steel in 2006," press release (Brussels: 22 January 2007).

2. Ibid.

3. Ibid.

4. "World Steel Output Hits Record in 2006," *Agence France-Presse*, 23 January 2007.

5. IISI, op. cit. note 1.

6. Ibid.

7. Ibid.

8. Iron and Steel Statistics Bureau (ISSB), "Steel Statistics in the News," at www.issb.co.uk/steel_news, viewed 21 February 2007.

9. ISSB, "The Race to Consolidate," at www.issb.co.uk, viewed 21 February 2007.

10. James Kanter, Heather Timmons, and Anand Giridharadas, "Arcelor Agrees to Mittal Takeover," *International Herald Tribune,* 25 June 2006.

11. Ibid.

12. Anuj Chopra, "India's Steel Industry Steps onto World Stage," *Christian Science Monitor*, 12 February 2007.

13. ISSB, op. cit. note 8.

14. Organisation for Economic Co-operation and Development (OECD), "OECD Steel Committee Sees Market Outlook Bright But Slower Demand Expected in 2007," press release (Paris: 8 November 2006).

15. "World 2006 Steel Demand Seen Rising 9 Percent—IISI," *Reuters*, 2 October 2006.

16. Kong Moon-kee, "East Asia Accounts for 54% of Global Steel Output," *Korea Times*, 14 February 2007.

17. "World 2006 Steel Demand," op. cit. note 15.

18. OECD, op. cit. note 14.

19. IISB, "The World's Top Trading Countries," at www.issb.co.uk, viewed 21 February 2007.

20. Embassy of the People's Republic of China, "China's Steel Exports Hit Record High Last Year," press release (Washington, DC: 10 January 2007).

21. IISB, op. cit. note 19.

22. Steel Recycling Institute, "Steel Recycling in the U.S. Continues Its Record Pace in 2005," press release (Washington, DC: 25 April 2006).

23. U.S. Geological Survey, *Mineral Commodity Summaries 2007* (Washington, DC: U.S.

Notes

Government Printing Office, January 2007).

24. Ibid.

25. Ibid.

26. IISI, "64.9% of Steel Cans Are Recycled," press release (Brussels: 10 January 2007).

27. Ibid.

ALUMINUM PRODUCTION CONTINUES UPWARD (pages 58–59)

1. U.S. Geological Survey (USGS), "Aluminum," *Mineral Commodity Summaries 2007* (Washington, DC: U.S. Government Printing Office (GPO), 2006), pp. 18–19.

2. Worldwatch calculation based on data in USGS, "Aluminum," *Mineral Commodity Summaries* (Washington, DC: GPO, various years).

3. USGS, op. cit. note 1, pp. 18–19.

4. Patricia Plunkert, aluminum specialist, USGS, Reston VA, e-mail to Andrew Wilkins, Worldwatch Institute, 3 December 2005.

5. USGS, "Aluminum Statistics and Information," at minerals.er.usgs.gov/minerals/pubs/commodity/aluminum/, viewed 28 February 2007.

6. Reuters Foundation, "FACTBOX—World Bauxite Reserves and Production," at www.alertnet.org/thenews/newsdesk/L15774125.htm, viewed 28 February 2007.

7. Ibid.

8. Worldwatch calculation based on data in USGS, op. cit. note 1, pp. 18–19.

9. Ibid.

10. "Aluminum and Bauxite," Mineral Information Institute, at www.mii.org/Minerals/photoal.html, viewed 28 February 2007.

11. "Emirates to Be Site of Largest Aluminum Plant," *International Herald Tribune*, 19 February 2006.

12. Worldwatch calculation based on data in USGS, op. cit. note 1, pp. 18–19.

13. Ibid.

14. Ibid.

15. Worldwatch calculation based on 1992 data in USGS, "Aluminum Statistics," in USGS Data Series 140, "Historical Statistics for Mineral and Material Commodities in the United States," at minerals.usgs.gov/ds/2005/140/, viewed 28 February 2007, and on 2006 data in USGS, op. cit. note 1, pp. 18–19.

16. USGS, op. cit. note 1, pp. 18–19.

17. The Aluminum Association, Inc., "Secondary Aluminum Smelters Feel Pinch," press release (Arlington, VA: 5 February 2004).

18. Worldwatch calculation based on total electricity from U.S. Department of Energy, Energy Information Administration, "Total Electric Power," at www.eia.doe.gov/emeu/international/electricitygeneration.html, and on International Aluminium Institute, "Electrical Power Used in Primary Aluminium Production," at www.world-aluminium.org/default.asp, viewed 2 March 2007.

19. Commonwealth Scientific and Industrial Research Organisation, "Aluminium Production: Reducing Costs and Energy Consumption," at www.csiro.au/science/ps18o.html#1, viewed 28 February 2007.

20. International Aluminium Institute, op. cit. note 18.

21. Ibid.

22. USGS, op. cit. note 1, pp. 18–19.

23. Ibid.

24. Subodh K. Das et al., "Energy Implications of the Changing World of Aluminum Metal Supply," *JOM*, August 2004, pp. 14–17.

25. Ibid.

26. Ibid.

27. Ibid.

28. International Aluminium Institute, "Aluminium Recycling," at www.world-aluminium.org/environment/recycling/index.html, viewed 2 March 2007.

GOLD MINING OUTPUT DROPS SLIGHTLY (pages 60–61)

1. GFMS, "Publication of Gold Survey 2006—Update 2, Mine Production," press release (London: 18 January 2007); historical data in Figure 1 from Earle Amey, commodity specialist, U.S. Geological Survey, letter to (10 January 2000) and discussion with (10 February 2000) Payal Sampat, Worldwatch Institute.

2. GFMS, op. cit. note 1.

3. Ibid.

4. Ibid.

5. Ibid.

6. Ibid.

7. Ibid.

8. China Gold Association, "Gold Industry's Performance in 2006" (Beijing: 31 January 2007).

9. Stockpile total based on data from World Gold

Council and GFMS.

10. Julius Baer Group, "Research News: The Lure of Gold" (Zurich: 30 October 2006).

11. Data from Kitco Inc. online database, at www.kit co.com/londonfix/gold.londonfix06.html, viewed 5 February 2007.

12. Ibid.; 25-year high based on earlier high price from Vikas Bajaj, "Gold Hits a 25-Year High," *New York Times*, 6 April 2006.

13. Data from "Gold Price Overview," US Gold Daily Quotes online, at www.usagold.com/DailyQuotes .html.

14. Julius Baer Group, op. cit. note 10.

15. GFMS, "Publication of Gold Survey 2006—Update 2, Official Sector," press release (London: 18 January 2007).

16. Ibid.

17. RBC Capital Markets, "Global Gold Outlook," *Mining Journal Online,* 1 November 2006.

18. GFMS, "Publication of Gold Survey 2006—Update 2, Main Highlights," press release (London: 18 January 2007).

19. Kevin Demeritt, "China's Gold Rush Triggers Skyrocketing Gold Prices," *Gold Central*, 15 January 2007.

20. GFMS, op. cit. note 17.

21. Adrian Ash, "Gold Goes All Google! Outlook for 2007," *Financial Sense,* 25 January 2007.

22. Ibid.

23. Ibid.

24. No Dirty Gold, "Community Voices," at www.no dirtygold.org.

25. Birks & Mayors Inc., "Birks & Mayors Endorses Boreal Forest Conservation—Canada's Premier Jewellery Retailer Calls for More Responsible Gold and Diamond Mining," *CNW Telbec*, 8 February 2007.

26. No Dirty Gold, "Retailers Who Support the Golden Rules," at www.nodirtygold.org, viewed 8 March 2007.

ROUNDWOOD PRODUCTION UP (pages 62–63)

1. U.N. Food and Agriculture Organization (FAO), *FAOSTAT Statistical Database*, at faostat.fao.org, updated 22 December 2006.

2. Worldwatch calculation based on data in FAO, op. cit. note 1.

3. Ibid.

4. Ibid.

5. Ibid.

6. U.N. Economic Commission for Europe (UNECE) and FAO, *Forest Products Annual Market Review 2005–2006*, (New York: United Nations, 2006), p. 5.

7. Ibid.

8. Worldwatch calculation based on data in FAO, op. cit. note 1.

9. UNECE and FAO, op. cit. note 6, p. 4.

10. Ibid., pp. 4, 19.

11. Ibid., p. 19.

12. Ibid., p. 4.

13. Ibid., pp. 87–95.

14. Ibid., pp. 23–24.

15. Ibid., p. 2.

16. World Wide Fund for Nature (WWF), *Failing the Forests: Europe's Illegal Timber Trade* (Surrey, U.K.: WWF-UK, 2004).

17. Ibid.

18. Greenpeace International, *Sharing the Blame: Global Consumption and China's Role in Ancient Forest Destruction* (Amsterdam: Greenpeace International and Greenpeace China: 2006).

19. UNECE and FAO, op. cit. note 6, p. 98.

20. Ibid., p. 101.

21. Ibid., p. 103.

22. Ibid., p. 101.

23. Ibid., p. 103.

24. Ibid., p. 102.

25. Ibid.

VEHICLE PRODUCTION RISES SHARPLY (pages 66–67)

1. Colin Couchman, Global Insight Automotive Group, e-mail to author, 11 January 2007.

2. Ibid.

3. Ibid.

4. Ibid.

5. Ibid.

6. Ibid.

7. Ward's Automotive Group, *Ward's Motor Vehicle Facts & Figures 2006* (Southfield, MI: 2006), p. 60.

8. Keith Bradsher, "Thanks to Detroit, China is Poised to Lead," *New York Times*, 12 March 2006.

9. Sales in 2005 from ibid.; 2005 passenger car fleet from Paul Zajac, Ward's Automotive Group, e-mail

to author, 31 January 2007.

10. Bradsher, op. cit. note 8.

11. Keith Bradsher, "China Seeking Auto Industry, Piece by Piece," *New York Times*, 17 February 2006.

12. Ward's Automotive Group, op. cit. note 7, p. 15.

13. Ibid.

14. Micheline Maynard, "Ford Chief Sees Small as Virtue and Necessity," *New York Times*, 26 January 2007.

15. Micheline Maynard and Martin Fackler, "Toyota is Poised to Supplant G.M. as World's Largest Carmaker," *New York Times*, 23 December 2006.

16. "Toyota to Boost Global Sales of Hybrid Cars," *People's Daily Online*, 25 January 2007.

17. Michael P. Walsh, "Cars Heavier and Faster, But U.S. Fuel Economy Unchanged," *Car Lines*, August 2006, p. 23.

18. Matthew L. Wald, "U.S. Raises Standards on Mileage," *New York Times*, 30 March 2006.

19. Michael P. Walsh, "Diesel Car Sales Seen Peaking in Europe," *Car Lines*, October 2006, p. 13.

20. Ibid., p. 14.

21. Ibid.

22. Michael P. Walsh, "EU Publishes Annual Report on CO_2 Emissions from New Cars," *Car Lines*, October 2006, p. 6.

23. Ibid., p. 5.

24. "Car Firms Facing Pollution Curbs," *BBC News Online*, 7 February 2007; Constant Brand, "EU Chief Backs Rules to Force Car Makers to Cut CO_2 Emissions," *Environmental News Network*, 23 January 2007.

25. "Canada Auto Union Alarmed by Tough Emissions Talk," *Environmental News Network*, 12 January 2007.

26. Michael P. Walsh, "California Sues Carmakers over Global Warming," *Car Lines*, October 2006, p. 31.

BICYCLE PRODUCTION UP SLIGHTLY (pages 68–69)

1. Global production numbers from United Nations, *The Growth of World Industry*, 1969 Edition, vol. II (New York: 1971), from United Nations, *Yearbook of Industrial Statistics*, 1979 and 1989 editions, vol. II (New York: 1981 and 1991), from *Interbike Directory*, various years, from United Nations, *Industrial Commodity Statistics Yearbook* (New York: various years), from *Bicycle Retailer and Industry News*, world market summary data, various years, and

from Otto Beaujon, managing editor, *Bike Europe Magazine*, e-mail to author, 10 March 2006.

2. "World Players in the Bicycle Market," in John Crenshaw, "China's Two-Wheeled Juggernaut Keeps Rolling Along," *Bicycle Retailer and Industry News*, 1 April 2006, p. 40.

3. Worldwatch calculation based on data cited in note 1.

4. "World Players in the Bicycle Market," op. cit. note 2.

5. "144 Percent: Mexico Doesn't Mess Around," *Bicycle Retailer and Industry News*, 1 November 2005, p. 47.

6. John Crenshaw, "EU Passes Anti-Dumping Duties on China, Vietnam," *Bicycle Retailer and Industry News*, 15 August 2005, p. 28.

7. Crenshaw, op. cit. note 2, p. 40.

8. "Indians Face News: Some Not So Good," *Bicycle Retailer and Industry News*, 15 March 2006, p. 28.

9. Crenshaw, op. cit. note 2, p. 40.

10. Doug McClellan, "E-Bike Makers Scramble for a Share of Swelling Market," *Bicycle Retailer and Industry News*, 1 April 2006, p. 49.

11. Doug McClellan, "New Standards May Boost E-Bike Market in Europe," *Bicycle Retailer and Industry News*, 1 November 2006, p. 32.

12. "Taiwan Builds Bike Paths, Promotes Cycling," *Bicycle Retailer and Industry News*, 15 July 2006, p. 27.

13. "Brits Take to Bikes, Infrastructure a Big Help," *Bicycle Retailer and Industry News*, 1 July 2006, p. 30.

14. "Aussie State Commits Big Bucks to Cycling," *Bicycle Retailer and Industry News*, 1 June 2006, p. 35.

15. "French Create National Cycling Czar Position," *Bicycle Retailer and Industry News*, 15 May 2006, p. 27.

16. Rachel Gordon, "Cycling Supporters on a Roll in S.F.," *San Francisco Chronicle*, 21 August 2006.

17. John Pucher and Ralph Buehler, "Why Canadians Cycle More than Americans: A Comparative Analysis of Bicycling Trends and Policies," *Journal of Transport Policy*, vol. 13 (2006), pp. 269–75.

18. Ibid.

19. Ibid.

20. Megan Tompkins, "Transportation Bill Delivers Estimated $4.5 Billion for Bicycling," *Bicycle Retailer and Industry News*, 1 April 2006, p. 38.

21. Ibid.

22. Robert Preer, "Start-up Offers Bike Rentals for $20, Targets College Students," *Bicycle Retailer and Indus-*

try News, 1 May 2006, p. 27.

23. Ibid.

AIR TRAVEL REACHES NEW HEIGHTS (pages 70–71)

1. International Civil Aviation Organization, "Development of World Scheduled Revenue Traffic," at www .ICAOdata.com. All data for 2005 are provisional.
2. Ibid.
3. Doug Gollan, *Elite Traveler Insider* (e-newsletter), 9 January and 6 February 2007.
4. General Aviation Manufacturers Association, *2006 General Aviation Statistical Databook* (Washington, DC: February 2007), p. 4.
5. "Gore vs. Hannity on Jet Emissions and Environment," *Helium Report,* 28 February 2007. Data provided by TerraPass.
6. International Air Transport Association (IATA), *Industry Times,* December 2006, p. 1.
7. Ibid.
8. Bennett Daviss, "Green Sky Thinking," *New Scientist,* 24 February 2007, p. 34.
9. Intergovernmental Panel on Climate Change, *Aviation and the Global Atmosphere: Summary for Policy Makers* (Cambridge, U.K.: Cambridge University Press, 1999).
10. European Union, "Questions & Answers on Aviation & Climate Change," press release (Brussels: 20 December 2006); Daviss, op. cit. note 8, p. 34.
11. IATA, "Debunking Some Persistent Myths About Air Transport and the Environment," at www.iata.org, no date.
12. IATA, "UN Guidelines on Emissions Trading Welcomed," press release (Montreal, PQ: 17 February 2007).
13. European Union, op. cit. note 10.
14. Ibid.
15. IATA, op. cit. note 12.
16. Daviss, op. cit. note 8, p. 35; tons to liters conversion from "Density Conversions," U.S. Environmental Protection Agency.
17. IATA, "Fuel Efficiency," at www.iata.org/whatwedo/ environment, viewed 16 February 2007.
18. Daviss, op. cit. note 8, pp. 36–37.
19. Peter Atkin, *Trash Landings: How Airlines and Airports Can Clean Up Their Recycling Programs* (New York: Natural Resources Defense Council, 2006), p. iv.
20. Ibid., p. v.

21. Centrair Japan International Airport Co., Ltd., *Centrair Green-Report 2006* (Tokoname, Aichi, Japan: 2006), p. 5.
22. Ibid., pp. 5, 17, 20.
23. Massachusetts Port Authority, "Logan Airport's Terminal A Awarded LEED Certification For Environmental Sustainability," press release (Boston: 2 August 2006).

CELL PHONES WIDELY USED, INTERNET GROWTH SLOWS (pages 72–73)

1. International Telecommunications Union (ITU), "Mobile Cellular, Subscribers per 100 People," at www.itu.int/ITU-D/ict/statistics, viewed 9 February 2007.
2. New subscribers is a Worldwatch calculation from ITU, op. cit. note 1; populations of United States and Canada from U.S. Bureau of the Census, *International Data Base,* electronic database, Suitland, MD, updated 24 August 2006.
3. Worldwatch calculation from ITU, op. cit. note 1.
4. ITU, "Main Telephone Lines, Subscribers per 100 People," at www.itu.int/ITU-D/icteye/Indicators/ Indicators.aspx#, viewed 9 February 2007.
5. ITU, e-mail to author, 13 February 2007.
6. Ibid.
7. Ibid.
8. "Burgers Paid for by Mobile Phone," *BBC News,* 27 February 2007.
9. Gail Nakada, "Japanese Use Cell Phone QR Bar Code Readers to Check Food Safety," *Wireless Watch Japan,* 14 May 2005.
10. David Zaks, "Text Messaging for Safe Water," *Worldchanging.com,* 27 October 2006.
11. Kim Hart, "Forecasts on Demand," *Washington Post,* 13 June 2006.
12. "Email, Weather, and Search Sites Are Most Popular Categories for Mobile Internet Use, According to Telephia," press release (San Francisco: Telephia, 7 September 2005).
13. Egil Juliussen, Computer Industry Almanac, e-mail to author, 12 February 2007; country data used to calculate world total are updated as available throughout the year.
14. ITU, "Internet Indicators: Hosts, Users, and Number of PCs," at www.itu.int/ITU-D/icteye/Indicators/ Indicators.aspx#, viewed 9 February 2007.

15. Ibid.; data are not available for 14 countries or territories, including Bhutan, Liberia, Myanmar, Nicaragua, and North Korea.

16. Slowest growth rate a Worldwatch calculation from Internet Systems Consortium, "ISC Domain Survey: Number of Internet Hosts," at www.isc.org, viewed 9 March 2007.

17. "That's What Little Notebooks Are Made Of," *Business Week*, 26 March 2007; Occupational Safety & Health Administration, U.S. Department of Labor, "Safety and Health Topics: Toxic Metals," at www.osha.gov/SLTC/metalsheavy/index.html.

18. Alan Deutschman, "There's Gold in Them Thar Smelly Hills," *Fast Company*, July 2006, p. 96.

19. Greenpeace China, *Recycling of Electronic Wastes in China and India: Workplace and Environmental Contamination* (Hong Kong: August 2005), p. 7.

20. Mure Dickie, "Google's 'Old Dog' Taught Chinese Tricks," *Financial Times,* 12 April 2006.

21. Jonathan Fildes, "Web Inventor Warns of 'Dark' Net," *BBC News,* 23 May 2006.

22. Ibid.

NUMBER OF VIOLENT CONFLICTS STEADY (pages 76–77)

1. Arbeitsgemeinschaft Kriegsursachenforschung (AKUF), "Zahl der kriegerischen Konflikte gegenüber dem Vorjahr unverändert," press release (Hamburg, Germany: University of Hamburg, 17 December 2006). AKUF data are continually being revised as improved information becomes available. Thus the numbers here differ from those reported in earlier editions of *Vital Signs*.

2. Ibid.

3. Ibid.

4. Ibid.

5. Ibid.

6. Ibid.

7. Nils Petter Gleditsch et al., "Armed Conflict: 1946–2001: A New Dataset," *Journal of Peace Research*, vol. 39, no. 5 (2002), pp. 615–37; updated information in Peace Research Institute Oslo, "The PRIO/Uppsala Armed Conflict Dataset. Armed Conflict, Version 4-2006b," at new.prio.no/CSCW-Datasets/Data-on-Armed-Conflict/Uppsala PRIO-Armed-Conflicts-Dataset/UppsalaPRIO-Armed -Conflict-Dataset/.

8. Heidelberger Institut für Internationale Konfliktforschung (HIIK), *Conflict Barometer 2006* (Heidelberg, Germany: Institute for Political Science, University of Heidelberg, 2006), p. 1; Julian-G. Albert, HIIK, e-mail to author, 4 January 2007.

9. Calculated from HIIK, op. cit. note 8.

10. Congo estimates from B. Coghlan et al., "Mortality in the Democratic Republic of Congo: A Nationwide Survey," *The Lancet*, 7 January 2006, pp. 44–51; Darfur estimates from "Q&A: Sudan's Darfur Conflict," *BBC News Online*, 29 January 2007.

11. Gilbert Burnham et al., "Mortality After the 2003 Invasion of Iraq: A Cross-Sectional Cluster Sample Survey," *The Lancet*, published online, 11 October 2006.

12. Sabrina Tavernise and Donald G. McNeil, Jr., "Iraqi Dead May Total 600,000, Study Says," *New York Times*, 11 October 2006; Sabrina Tavernise, "Iraqi Death Toll Exceeded 34,000 in 2006, U.N. Says," *New York Times*, 17 January 2007.

13. U.N. High Commissioner for Refugees, "Refugees by Numbers 2006 edition," at www.unhcr.org/cgi-bin/texis/vtx/basics/opendoc.htm?tbl=BASICS&id=3b02 8097c, viewed 4 January 2007.

14. Ibid.

15. Calculated from U.S. Committee for Refugees and Immigrants, *World Refugee Survey 2006* (Washington, DC: 2006), Table 9.

16. Internal Displacement Monitoring Centre, "Global Statistics," at www.internal-displacement.org, viewed 4 January 2007.

17. Ibid., viewed 13 February 2007.

18. HIIK, op. cit. note 8, p. 6.

19. Ibid., p. 7.

20. See "Peacekeeping Hits New Record" in this volume.

21. HIIK, op. cit. note 8, p. 7.

PEACEKEEPING EXPENDITURES HIT NEW RECORD (pages 78–79)

1. U.N. Department of Public Information (UNDPI), "United Nations Peacekeeping Operations. Background Note" (New York: 31 December 2006, and earlier editions); Worldwatch database. All dollar amounts are in 2006 dollars.

2. U.N. Department of Peacekeeping Operations (UNDPKO), "Monthly Summary of Contributors,"

at www.un.org/Depts/dpko/dpko/contributors/index.htm, viewed 2 January 2007; personnel number also based on William Durch, Henry L. Stimson Center, Washington, DC, e-mail to author, 9 January 1996, and on Global Policy Forum, at www.globalpolicy.org/security/peacekpg/data/pkomctab.htm, viewed 2 January 2007.

3. UNDPI, op. cit. note 1; UNDPI, "United Nations Political and Peace-Building Missions. Background Note" (New York: 31 October 2006). Two of these assistance missions—in Afghanistan and Sierra Leone—are directed by the UNDPKO; the others, by the U.N. Department of Political Affairs.

4. "Twenty Days in August: The Security Council Sets Massive New Challenges for UN Peacekeeping," Security Council Report, No. 5, 8 September 2006; Colum Lynch, "Peacekeeping Grows, Strains U.N.," Washington Post, 17 September 2006.

5. Mark Turner, "UN Fears Overstretch," Financial Times, 22 August 2006.

6. Mark Turner, "UN Snubs Calls to Shield Refugees," Financial Times, 30 December 2006; Mark Turner, "UN Seeks Force for Central Africa," Financial Times, 21 February 2007.

7. Petter Stålenheim et al., "Military Expenditure," in Stockholm International Peace Research Institute (SIPRI), SIPRI Yearbook 2006 (New York: Oxford University Press, 2006), p. 295.

8. UNDPI, op. cit. note 1.

9. Author's calculation, based on data from UNDPKO, op. cit. note 2. The percentage figures in this and the following paragraph refer to peacekeeping personnel excluding civilian staff.

10. Ibid.

11. Calculated from UNDPKO, op. cit. note 2.

12. UNDPI, op. cit. note 1.

13. Ibid.

14. Ibid.

15. Ibid.

16. "U.N. Prepares to Start a New Peacekeeping Mission in East Timor," New York Times, 14 June 2006.

17. UNDPKO, "ONUB: United Nations Operation in Burundi," at www.un.org/Depts/dpko/missions/onub/; leaving ahead of schedule from Richard Gowan, "The UN and Peacekeeping: Taking the Strain?" Signal, autumn 2006, p. 50.

18. Global Policy Forum, "US vs. Total Debt to the UN: 2006," at www.globalpolicy.org/finance/tables/core/un-us-06.htm, viewed 28 December 2006.

19. Global Policy Forum, "Contributions Owing to the UN for Peacekeeping Operations: 2006," at www.globalpolicy.org/finance/tables/pko/due2006.htm, viewed 2 January 2007.

20. Ibid.

21. Worldwatch Institute database.

22. Calculated from Sharon Wiharta, "Multilateral Peace Missions in 2005," in SIPRI, op. cit. note 7, pp. 158–93, from "Peace, Peacebuilding and Crisis Prevention Missions 2006," Center for International Peace Operations, Berlin, Germany, www.zif-berlin.org, and from Future of Peace Operations Program, "Numbers of Uniformed Personnel in Peace Operations at Mid-Year, 1948–2006," undated, supplemental material to William J. Durch and Tobias C. Berkman, Who Should Keep the Peace? Providing Security for Twenty-First-Century Peace Operations (Washington, DC: Henry L. Stimson Center, 2006).

23. Worldwatch Institute database.

24. Close to 90 percent of all personnel in U.N.-led peacekeeping operations in 2006 served in missions that were authorized under Chapter 7. In 1993, the share was just 38 percent. Future of Peace Operations Program, op. cit. note 22.

NUCLEAR WEAPONS TREATY ERODING
(pages 80–81)

1. Robert S. Norris and Hans M. Kristensen, "Global Nuclear Stockpiles, 1945–2006," Bulletin of the Atomic Scientists, July/August 2006, pp. 64–66. Warhead data are estimates because a great deal of information about these arsenals remains a government secret. Thus the reported numbers are revised over time as better information becomes available.

2. Carnegie Endowment for International Peace, "Nuclear Numbers," at www.carnegieendowment.org/npp/numbers/default.cfm, viewed 7 January 2007.

3. Norris and Kristensen, op. cit. note 1.

4. Carnegie Endowment for International Peace, op. cit. note 2.

5. Ibid.

6. Norris and Kristensen, op. cit. note 1.

7. Ibid.

8. Shannon N. Kile, Vitaly Fedchenko, and Hans M. Kristensen, "World Nuclear Forces, 2006," in Stock-

Notes

holm International Peace Research Institute, *SIPRI Yearbook 2006: Armaments, Disarmament and International Security* (New York: Oxford University Press, 2006), p. 640.

9. Norris and Kristensen, op. cit. note 1.

10. Ibid.

11. Ibid.

12. "Treaty on the Non-proliferation of Nuclear Weapons," 5 March 1970 (entry into force), text at United Nations, disarmament.un.org/wmd/npt/npttext.html.

13. Kile, Fedchenko, and Kristensen, op. cit. note 8, p. 639; William J. Broad, "U.S. Has Plans to Again Make Own Plutonium," *New York Times*, 27 June 2005; Walter Pincus, "Nuclear Officials Seek Approval for Warhead," *Washington Post*, 7 February 2007.

14. William J. Broad, David E. Sanger, and Tom Shanker, "U.S. Selecting Hybrid Design for Warheads," *New York Times*, 7 January 2007.

15. Kile, Fedchenko, and Kristensen, op. cit. note 8, p. 639.

16. Ibid., p. 640.

17. Ibid.

18. Alan Cowell, "Blair Urges Keeping Nuclear Arms Program Alive," *New York Times*, 5 December 2006.

19. Ibid.

20. Lower estimate from Norris and Kristensen, op. cit. note 1; higher fissile material estimate from Carnegie Endowment for International Peace, op. cit. note 2.

21. Norris and Kristensen, op. cit. note 1.

22. Carnegie Endowment for International Peace, "Korea," at www.carnegieendowment.org/npp/country/index.cfm?fa=view&id=1000090.

23. Jill Marie Parillo, "Iran's Nuclear Program," *Carnegie Fact Sheet* (Carnegie Endowment for International Peace), September 2006.

24. United Nations, Security Council Resolution 1696, 31 July 2006; United Nations, Security Council Resolution 1737, 23 December 2006.

25. See, for instance, Seymour M. Hersh, "The Iran Plans," *The New Yorker*, 17 April 2006, and Seymour M. Hersh, "The Next Act," *The New Yorker*, 27 November 2006; "Target Iran—Air Strikes," GlobalSecurity.org; Gabriel Ronay, "America 'Poised to Strike at Iran's Nuclear Sites' from Bases in Bulgaria and Romania," *Sunday Herald* (Scotland), 28 January 2007.

26. "U.S., India Reach Deal On Nuclear Cooperation," *Washington Post*, 3 March 2006.

AGRIBUSINESSES CONSOLIDATE POWER (pages 86–87)

1. ETC Group, *Oligopoly, Inc. 2005: Concentration in Corporate Power*, Communiqué Issue #91 (Ottawa, ON: December 2005).

2. Pew Initiative on Food and Biotechnology, "Factsheet: Genetically Modified Crops in the United States," Washington, DC, August 2004.

3. Sophia Murphy, *Concentrated Market Power and Agricultural Trade*, EcoFair Trade Dialog Discussion Paper No. 1 (Berlin: Heinrich Böll Foundation, August 2006), p. 10.

4. C. S. Srinivasan, "Concentration in Ownership of Plant Variety Rights: Some Implications for Developing Countries," *Food Policy*, October–December 2003, pp. 519–46.

5. Keith Collins, *Statement before the Senate Committee on Appropriations* (Washington, DC: U.S. Government Printing Office (GPO), 2001); Democratic Staff Committee on Agriculture, Nutrition and Forestry, *Economic Concentration and Structural Change in the Food and Agriculture Sector: Trends, Consequences and Policy Options* (Washington, DC: GPO, 2004).

6. Mary Hendrickson and William Heffernan, *Concentration of Agricultural Markets* (Columbia, MO: Department of Rural Sociology, University of Missouri, 2005).

7. Murphy, op. cit. note 3, p. 9.

8. Ibid.

9. William Heffernan, "Social Consequences of Factory Hog Production Systems," in *Understanding the Impacts of Large-scale Swine Production: Proceedings from an Interdisciplinary Scientific Workshop* (Des Moines, IA: 2005); M. S. Honeyman, "Sustainability Issues of U.S. Swine Production," *Journal of Animal Science*, June 1996, pp. 1410–17.

10. Robert Kemp, *Innovation in the Livestock Industry: Implications for Animal Genetic Improvement Programs* (Ottawa, ON: Canadian Biotechnology Advisory Committee, 2001); Harvey Blackburn et al., *United States of America Country Report for FAO's State of the World's Animal Genetic Resources* (Washington, DC:

U.S. Department of Agriculture (USDA), 2003); Canadian Farm Animal Genetic Resources Foundation, *The Need for an Animal Genetic Resource Policy for Canada* (Brighton, ON: 2003).

11. Janet Raloff, "Dying Breeds," *Science News*, 4 October 1997; Harlan Ritchie, *Where is the Beef Seedstock Industry Headed?* (East Lansing, MI: Michigan State University, 2002).

12. Honeyman, op. cit. note 9; David Pimentel et al., "Economic and Environmental Benefits of Biodiversity," *BioScience*, December 1997, pp. 747–57; D. R. Notter, "The Importance of Genetic Diversity in Livestock Populations of the Future," *Journal of Animal Science*, January 1999, pp. 61–69.

13. Agricultural Marketing Service, *Contracting in Agriculture: Making the Right Decision* (Washington, DC: USDA, 2006); Neil Harl, "Contract Agriculture: Will it Tip the Balance?" *Leopold Letter* (Ames, IA: Iowa State University, 1998); Collins, op. cit. note 5; Clare Hinrichs and R. Welsh, "The Effects of the Industrialization of US Livestock Agriculture on Promoting Sustainable Production Practices," *Agriculture and Human Values*, June 2003, pp. 125–41.

14. USDA, *Structure and Finances of U.S. Farms: 2005 Family Farm Report* (Washington, DC: 2005).

15. Oli Brown, *Supermarket Buying Power, Global Commodity Chains and Smallholder Farmers in the Developing World*, Human Development Report Office Occasional Paper (New York: U.N. Development Programme, 2005).

16. J. Wilkinson and G. Flexor, *Brazilian Agrofood, Transnationalization and Market Concentration* (Rio de Janiero: Rural Federal University, 2005).

17. Hendrickson and Heffernan, op. cit. note 6.

18. Christopher Delgado et al., "Livestock to 2020: The Next Food Revolution," *Outlook on Agriculture*, March 2001, pp. 27–29.

19. Myriam Vander Stichele, Sanne van der Wal, and Joris Oldenziel, *Who Reaps the Fruit* (Amsterdam: SOMO, 2005), p. 49.

20. USDA Economic Research Service, Food Price Spread Briefing Room, at www.ers.usda.gov/Briefing/FoodPriceSpreads, viewed 20 February 2007.

21. National Farmers Union, *The Farm Crisis, Bigger Farms, and the Myths of "Competition" and "Efficiency"* (Saskatoon, SK: 2003), p. 12.

22. See, for example, Emek Basker, *Selling a Cheaper Mousetrap: Wal-Mart's Effect on Retail Prices* (Columbia, MO: University of Missouri, 2005).

23. Lawrence Weiss, ed., *Concentration and Price* (Cambridge, MA: The MIT Press, 1989), pp. 266–83; Bruce Marion et al., "Strategic Groups, Competition and Retail Food Prices," in Ronald Cotterill, ed., *Competitive Strategy Analysis in the Food System* (Boulder, CO: Westview Press, 1993), p. 197; Ronald Cotterill, "Measuring Market Power in the Demsetz Quality Critique in the Retail Food Industry," *Agribusiness*, vol. 101, no. 15 (1999); Peter Carstensen, "Concentration and the Destruction of Competition in Agricultural Markets: The Case for Change in Public Policy," *Wisconsin Law Review*, spring 2000, p. 531.

24. Carstensen, op. cit. note 23.

25. Democratic Staff Committee, op. cit. note 5.

26. Ibid.

27. Ibid.

28. General Accounting Office, *Packers and Stockyards Programs: Actions Needed to Improve Investigations of Competitive Practices* (Washington, DC: 2000).

29. Tescopoly, at www.tescopoly.org, viewed 9 March 2007.

30. Wal-Mart Watch, at walmartwatch.com/issues/labor_relations, viewed 9 March 2007; Wal-Mart Class Action, at www.walmartclass.com/public_home.html, viewed 9 March 2007.

EGG PRODUCTION DOUBLES SINCE 1990 (pages 88–89)

1. U.N. Food and Agriculture Organization (FAO), *FAOSTAT Statistical Database*, at faostat.fao.org, updated 2006.

2. Ibid.

3. International Egg Commission, "The World Egg Industry: A Few Facts and Figures," at www.internationalegg.com/corporate/eggindustry, viewed 27 February 2007.

4. Dr. Edward Gillin, *World Egg and Poultry Meat Production, Trade, and Supply Present and the Future* (Rome: FAO, 2001).

5. Ibid.; FAO, op. cit. note 1.

6. FAO, op. cit. note 1.

7. Ibid.

8. Hongge Wang, "The Chinese Poultry Industry at a Glance." *World Poultry*, vol. 22, no. 8 (2006), pp. 10–11.

9. Ibid.

Notes

10. FAO, op. cit. note 1.
11. Ibid.
12. Gillin, op. cit. note 4.
13. Ibid.
14. Ibid.
15. Ibid.
16. Ibid.
17. Ibid.
18. Ibid.
19. Wang, op. cit. note 8; "China Egg Production Gets into Full Swing,"*AP-Foodtechnology.com*, 7 February 2004; Access Asia, *Eggs in China*, updated 1 August 2004.
20. Compassion in World Farming, at www.ciwf.org.uk/campaigns/primary_campaigns/egg-laying.html, viewed 9 April 2007.
21. Wang, op. cit. note 8.
22. Weibe van der Sluis, "Major Changes in the Asian Egg Markets," *World Poultry*, vol. 21, no.1 (2005), pp. 10–11.
23. Gillin, op. cit. note 4; "China Egg Production Gets into Full Swing," op. cit. note 19; Access Asia, op. cit. note 19.
24. Wang, op. cit. note 8.
25. Ibid.; Soren S. Kjaer, *Analyses of Sector Policies with Relevance Intensive Livestock Production Working Paper: The Case of Jiangsu Province, People's Republic of China: A Livestock Environment & Development (LEAD) Initiative* (Copenhagen: Danish Ministry of Environment and Energy, Nature and Forest Agency, 2001), p. 3.
26. World Society for the Protection of Animals, "Pollution by Manure, Slurry and Animal Carcases," at wspafarmwatch.org, viewed 27 February 2007.
27. "New Program to Clean Major Freshwater Lake in East China," *People's Daily*, 27 March 2002; H. Y. Guo, X. R. Wang, and J.G. Zhu, "Quantification and Index of Non-Point Source Pollution in Taihu Lake Region with GIS," *Environmental Geochemistry and Health*, June 2004, pp. 147–56.
28. Weibe van der Sluis and Joanne McIntyre, "Huge Variations in International Egg Costs," *World Poultry*, vol. 22, no. 1 (2006), p. 1; van der Sluis, op. cit. note 22; "China Egg Production Gets into Full Swing," op. cit. note 19; Access Asia, op. cit. note 19.
29. Wang, op. cit. note 8.
30. Van der Sluis, op. cit. note 22.
31. C. L. Delgado and Claire A. Narrod, *Policy, Technical, and Environmental Determinants and Implications of the Scaling-up of Livestock Production in Four Fast-growing Developing Countries: A Synthesis* (Washington, DC: International Food Policy Research Institute, 2003).
32. FAO, "Animal Health Special Report: Avian Influenza, Questions & Answers," at www.fao.org/ag/againfo/subjects/en/health/diseases-cards/avian _qa.html#1, viewed 26 February 2007.
33. Humane Society of the United States, *An HSUS Report: The Welfare of Animals in the Egg Industry* (Washington, DC: February 2006).
34. "EU Bans Battery Hen Cages," *BBC News Online*, 28 January 1999.
35. Cage-free egg trend victory page, Humane Society of the United States, at www.hsus.org/farm/camp/victories.html, viewed 27 February 2007.
36. Joyce D'Silva, "Asia for Animals," *Farm Animal Voice: The Magazine of Compassion in World Farming*, spring 2007, p. 13.
37. Ibid.

AVIAN FLU SPREADS (pages 90–91)

1. World Health Organization (WHO), "Cumulative Number of Confirmed Human Cases of Avian Influenza A/(H5N1) Reported to WHO" (Geneva: 2 April 2007); Juan Lubroth, Head, Infectious Disease Group/EMP RES, and Senior Officer, Animal Health Service, Food and Agriculture Organization (FAO), e-mail to author, 4 January 2007.
2. WHO, op. cit. note 1; Lubroth, op. cit. note 1.
3. WHO, "Ten Things You Need to Know About Pandemic Influenza" (Geneva: 14 October 2005).
4. Declan Butler and Jaqueline Ruttiman, "Avian Flu and the New World," *News@Nature.com*, 10 May 2006; "Global Impact of Bird Flu," *BBC World News*, updated January 2007; Table 1 from WHO, op. cit. note 1.
5. Butler and Ruttiman, op. cit. note 4.
6. FAO, Animal Production and Health Division, "Avian Influenza," at www.fao.org/ag/againfo/subjects/en/health/diseases-cards/special_avian.html.
7. FAO, *FAOSTAT Statistical Database*, at faostat.fao.org, viewed January 2007.
8. FAO, op. cit. note 6.
9. WHO, "Avian Influenza Frequently Asked Questions," at www.who.int/csr/disease/avian _influenza/avian_faqs/en/index.html#whatis.
10. Ibid.
11. Ibid.

12. C. J. Murray et al., "Estimation of Potential Global Pandemic Influenza Mortality on the Basis of Vital Registry Data from the 1918–20 Pandemic: A Quantitative Analysis," *The Lancet*, 23 December 2006, pp. 2211–18.

13. WHO, op. cit. note 3.

14. Ibid.

15. World Bank, East Asia and Pacific Region, "Economic Impact of Avian Flu," at web.worldbank .org/WBSITE/EXTERNAL/COUNTRIES/EASTASIA PACIFICEXT/EXTEAPREGTOPHEANUT/0,,content MDK:20713527~pagePK:34004173~piPK:3400370 7~theSitePK:503048,00.html.

16. World Bank, East Asia and Pacific Region, "Avian Flu: Economic Losses Could Top US$800 Billion" (Washington, DC: 8 November 2005); Jason Gale and Kristen Hallam, "Severe Flu Pandemic May Cost Up to $2 Trillion, World Bank Says," *Bloomberg.com*, 29 June 2006.

17. WHO, op. cit. note 9.

18. Ibid.

19. World Bank, *Avian and Human Influenza: Update on Financing Needs and Framework, Draft* (Washington, DC: 30 November 2006), p. 5.

20. GRAIN, *Fowl Play: The Poultry Industry's Central Role in the Bird Flu Crisis*, GRAIN Briefing (Barcelona, Spain: February 2006), p. 4.

21. Ibid.

22. See Christopher L. Delgado and Claire A. Narrod, *Impact of Changing Market Forces and Policies on Structural Change in the Livestock Industries of Selected Fast-Growing Developing Countries, Final Research Report on Phase I—Project on Livestock Industrialization, Trade, and Social-Health-Environmental Impacts in Developing Countries* (Rome: International Food Policy Research Institute and FAO, 2002), and Henning Steinfeld et al., *Livestock's Long Shadow, Environmental Issues and Options* (Rome: FAO, 2006) for more information on the growth of factory farming.

23. Michael Greger, *Bird Flu: A Virus of Our Own Hatching* (New York: Lantern Books, 2006), p. 109; Council for Agricultural Science and Technology, *Global Risks of Infectious Animal Diseases*, Issue Paper No. 28 (Ames, IA: February 2005).

24. Greger, op. cit. note 23.

25. FAO, Animal Production and Health Division, "Avian Influenza—Questions and Answers," *Animal Health Special Report*, at www.fao.org/ag/AGAInfo/ subjects/en/health/diseases-cards/avian_qa.html#1.

26. GRAIN, op. cit. note 20, p. 9.

27. "Deadly Bird Flu Found in Africa," *BBC News*, 8 February 2006.

28. GRAIN, op. cit. note 20, p. 17.

29. Christopher L. Delgado, Claire A. Narrod, and Marites M. Tiongco, Project on Livestock, Trade and Social-Health-Environmental Impacts in Developing Countries, *Policy, Technical, and Environmental Determinants and Implications of the Scaling-Up of Livestock Production in Four Fast-Growing Developing Countries: A Synthesis, Final Research Report of Phase II* (Rome: FAO, 2003); GRAIN, op. cit. note 20, p. 16.

30. Greger, op. cit. note 23, p. 33.

31. Michael Greger, e-mail to author, January 2007; Greger, op. cit. note 23, pp. 194–96.

32. GRAIN, op. cit. note 20.

33. Ibid.

34. Dr. Isabella Donatelli, "Serological Analysis of Serum Samples of Humans Exposed to Avian H7 Influenza Viruses in Italy Between 1999 and 2003," *Journal of Infectious Diseases*, 15 October 2005, p. 5; "Less Virulent Strains of Avian Influenza Can Infect Humans," press release (Alexandria, VA: Infectious Diseases Society of America, 13 September 2005).

35. GRAIN, op. cit. note 20.

36. Ibid.

CLIMATE CHANGE AFFECTS TERRESTRIAL BIODIVERSITY (pages 94–95)

1. Intergovernmental Panel on Climate Change (IPCC), *Climate Change 2007: The Physical Science Basis—Summary for Policymakers* (Geneva: 2007).

2. Millennium Ecosystem Assessment (MA), *Ecosystems and Human Well-being: Biodiversity Synthesis* (Washington, DC: World Resources Institute, 2005), p. 9.

3. A. Menzel and V. Dose, "Analysis of Long-term Time-series of Beginning of Flowering by Bayesian Function Estimation," *Meteorologische Zeitschrift*, June 2005, pp. 429–34.

4. D. R. Cayan et al., "Changes in the Onset of Spring in the Western United States," *Bulletin of the American Meteorological Society*, March 2001, pp. 399–415.

5. M. A. White, S. W. Running, and P. E. Thornton, "The Impact of Growing-season Length Variability

Notes

on Carbon Assimilation and Evapotranspiration over 88 Years in the Eastern US Deciduous Forest," *International Journal of Biometeorology*, February 1999, pp. 134–45.

6 . D. B. Roy and T. H. Sparks, "Phenology of British Butterflies and Climate Change," *Global Change Biology*, April 2000, pp. 407–16.

7. C. Stefanescu, J. Penuelas, and I. Filella, "Effects of Climate Change on the Phenology of Butterflies in the Northwest Mediterranean Basin," *Global Change Biology*, October 2003, p. 1494; M. L. Forister and A. M. Shapiro, "Climatic Trends and Advancing Spring Flight of Butterflies in Lowland California," *Global Change Biology*, July 2003, pp. 1130–35.

8. J. P. Gibbs and A. R. Breisch, "Climate Warming and Calling Phenology of Frogs near Ithaca, New York, 1900–1999," *Conservation Biology*, August 2001, pp. 1175–78; O. Hppop and K. Hppop, "North Atlantic Oscillation and Timing of Spring Migration in Birds," *Proceedings of the Royal Society B*, 7 February 2003, pp. 233–40.

9. H. Q. Crick et al., "UK Birds Are Laying Eggs Earlier" (letter), *Nature*, 7 August 1997, p. 526; P. O. Dunn and D. W. Winkler, "Climate Change Has Affected the Breeding Date of Tree Swallows Throughout North America," *Proceedings of the Royal Society B*, 22 December 1999, pp. 2487–90.

10. C. Both et al., "Large-scale Geographical Variation Confirms that Climate Change Causes Birds to Lay Earlier," *Proceedings of the Royal Society B*, 22 August 2004, pp. 1657–62.

11. C. Parmesan, "Ecological and Evolutionary Responses to Recent Climate Change," *Annual Review of Ecology, Evolution, and Systematics*, December 2006, pp. 637–69.

12. Ibid.

13. C. D. Thomas and J. J. Lennon, "Birds Extend Their Ranges Northwards," *Nature*, 20 May 1999, p. 213.

14. R. Hickling et al., "A Northward Shift of Range Margins in British Odonata," *Global Change Biology*, March 2005, pp. 502–06.

15. J. W. Wilson et al., "Changes to the Elevational Limits and Extent of Species Ranges Associated with Climate Change," *Ecology Letters*, November 2005, pp. 1138–46.

16. C. Parmesan and H. Galbraith, *Observed Ecological Impacts of Climate Change in North America* (Arlington, VA: Pew Center on Global Climate Change,

2004).

17. IPCC, op. cit. note 1; MA, op. cit. note 2.

18. Parmesan, op. cit. note 11.

19. M. E. Visser and C. Both, "Review. Shifts in Phenology due to Global Climate Change: The Need for a Yardstick," *Proceedings of the Royal Society B*, 22 December 2005, pp. 2561–69.

20. Reviewed in Parmesan, op. cit. note 11.

21. IPCC, *Climate Change 2007: Climate Change Impacts, Adaptation and Vulnerability—Summary for Policymakers* (Geneva: 2007), p. 3.

22. Ibid., p. 4.

23. Ibid., p. 8.

24. Ibid., p. 18.

25. Thomas Lovejoy, testimony before the subcommittee hearing on Global Warming and Wildlife, Environment and Public Works Committee, U.S. Senate, Washington, DC, 7 February 2007.

THREATS TO SPECIES ACCELERATE (pages 96–97)

1. IUCN Species Survival Commission (SSC), *2006 IUCN Red List of Threatened Species* (Gland, Switzerland, and Cambridge, U.K.: IUCN–The World Conservation Union, 2006).

2. Ibid.

3. Ibid.

4. Jonathan E. M. Baillie, Craig Hilton-Taylor, and Simon N. Stuart, eds., *2004 IUCN Red List of Threatened Species: A Global Species Assessment* (Gland, Switzerland, and Cambridge, U.K.: IUCN, 2004).

5. Arctic Climate Impact Assessment, *Impacts of a Warming Arctic* (Cambridge, U.K.: Cambridge University Press, 2004).

6. Andrew E. Derocher, Nicholas J. Lunn, and Ian Stirling, "Polar Bears in a Warming Climate," *Integrative & Comparative Biology*, April 2004, pp. 163–76.

7. IUCN SSC, op. cit. note 1.

8. Ibid.

9. Ibid.

10. Ibid.

11. Ibid.

12. Kevin G. Smith and William R. T. Darwall, compilers, *The Status and Distribution of Freshwater Fish Endemic to the Mediterranean Basin* (Gland, Switzerland: IUCN, 2006).

13. Ibid.

14. IUCN SSC, op. cit. note 1.

15. A. K. Hillman Smith et al., *Virunga National Park—North Aerial Census of March 2003* (publisher unknown).

16. Ibid.

17. IUCN SSC, op. cit. note 1.

18. BirdLife International, Species Factsheet: *Haliaeetus albicilla*, 2006, at www.birdlife.org, viewed 15 March 2007.

19. Ibid.

20. BirdLife International, *Threatened Birds of the World 2006* (Barcelona: Lynx Edicions, 2006).

21. Ibid.

22. Mekong Wetlands Biodiversity Conservation and Sustainable Use Programme, "Historic Agreement Signed to Save the Critically Endangered Migrating Giants of the Mekong…. But Is It Enough?" (Vientiane, Laos: 2006); Royal Society for the Protection of Birds, "Ray of Hope for Vultures Facing Extinction," press release (United Kingdom: 31 January 2006).

INVASIVE SPECIES DRIVE BIODIVERSITY LOSS (pages 98–99)

1. Millennium Ecosystem Assessment, *Ecosystems and Human Well-being: Biodiversity Synthesis* (Washington, DC: World Resources Institute, 2005), p. 42.

2. Ibid.

3. S. Lowe et al., *100 of the World's Worst Invasive Alien Species: A Selection from the Global Invasive Species Database* (Auckland, New Zealand: Invasive Species Specialist Group (ISSG), Species Survival Commission (SSC), IUCN–The World Conservation Union, 2000).

4. BirdLife International, "*Megapodius laperouse*" and "*Anas nesiotis*," in IUCN SSC, *2006 IUCN Red List of Threatened Species* (Gland, Switzerland, and Cambridge, U.K.: IUCN, 2006); G. J. Wiles et al., "Impacts of the Brown Tree Snake: Patterns of Decline and Species Persistence in Guam's Avifauna." *Conservation Biology*, vol. 17, no. 5 (2003); IUCN, Conservation International, and NatureServe, *Global Amphibian Assessment: 2006 Update*, at www.globalamphibians.org/update.htm; BirdLife International, "*Oxyura leucocephala*," in IUCN SSC, op. cit. this note; David M. Richardson and Brian W. Van Wilgen, "Invasive Alien Plants in South Africa: How Well Do We Understand the Ecological Im-

pacts?" *South African Journal of Science*, January/February 2004, pp. 45–52.

5. Lawrence R. Walker and Peter M. Vitousek, "An Invader Alters Germination and Growth of a Native Dominant Tree in Hawai'i," *Ecology*, August 1991, pp. 1449–55; Richardson and Van Wilgen, op. cit. note 4.

6. David Pimentel, Rodolfo Zuniga, and Doug Morrison, "Update on the Environmental and Economic Costs Associated with Alien-Invasive Species in the United States," *Ecological Economics*, 15 February 2005, pp. 273–88.

7. BirdLife International, *State of the World's Birds 2004: Indicators for Our Changing World* (Cambridge, U.K.: 2004).

8. Ibid.; Jonathan E. M. Baillie, Craig Hilton-Taylor, and Simon N. Stuart, eds., *2004 IUCN Red List of Threatened Species: A Global Species Assessment* (Gland, Switzerland, and Cambridge, U.K.: IUCN, 2004).

9. Baillie, Hilton-Taylor, and Stuart, op. cit. note 8.

10. BirdLife International, "*Anas nesiotis*," op. cit. note 4; I. C. S. Van Riper, "The Impact of Introduced Vectors and Avian Malaria on Insular Passeriform Bird Populations in Hawaii," *Bulletin of the Society of Vector Ecologists*, vol. 16, no. 1 (1991), pp. 59–83.

11. Walker and Vitousek, op. cit. note 5.

12. P. M. Vitousek, "Biological Invasions and Ecosystem Processes Towards an Integration of Population Biology and Ecosystem Studies," *Oikos*, vol. 57, no. 1 (1990), pp. 7–13.; Carla M. D'Antonio and Michelle Mack, "Exotic Grasses Potentially Slow Invasion of an N-fixing Tree into a Hawaiian Woodland," *Biological Invasions*, March 2001, pp. 69–73.

13. Richardson and Van Wilgen, op. cit. note 4.

14. Ibid.

15. Office of Technology Assessment, *Harmful Non-Indigenous Species in the United States* (Washington, DC: U.S. Government Printing Office, 1993).

16. David Pimentel et al., "Economic and Environmental Threats of Alien Plant, Animal, and Microbe Invasions," *Agriculture, Ecosystems & Environment*, March 2001, pp. 1–20.

17. See www.issg.org/database, www.nobanis.org, www.hear.org/pier, and nas.er.usgs.gov.

18. See www.gisinetwork.org.

19. N. F. Kümpel, and J. E. M. Baillie, "Options for a Global Indicator on Trends in Invasive Alien

Notes

Species," Report to the Secretariat of the Convention on Biological Diversity, Indicators and Assessments Unit (London: Institute of Zoology, Zoological Society of London, and SSC of IUCN, 2007). The GISD is available online at www.issg.org/database or in a CD-ROM.

20. "About The GISD," Global Invasive Species Database, at www.issg.org/database/welcome/aboutGISD.asp, viewed 21 March 2007.

21. Ibid.

OCEAN POLLUTION WORSENS AND SPREADS
(pages 100–01)

1. B. Worm et al., "Impacts of Biodiversity Loss on Ocean Ecosystem Services," *Science*, 3 November 2006, pp. 787–90; C. Nellemann and E. Corcoran, eds., *Our Precious Coasts: Marine Pollution, Climate Change and the Resilience of Coastal Ecosystems* (Arendal, Norway: U.N. Environment Programme, 2006). Pollution hotspots in Figure 1 from the following: Nellemann and Corcoran, op. cit. this note; J. Michael Beman, Kevin R. Arrigo, and Pamela A. Matson, "Agricultural Runoff Fuels Large Phytoplankton Blooms in Vulnerable Areas of the Ocean," *Nature*, 10 March 2005, pp. 211–14; Kenneth Weiss, "Plague of Plastic Chokes the Seas," *Los Angeles Times*, 2 August 2006; Oceana Web site, at www.oceana.org.

2. Nellemann and Corcoran, op. cit. note 1, p. 5.

3. Ibid.

4. Ibid., p. 16.

5. Woods Hole Oceanographic Institution, "Domestic Pollution and Sewage," at www.whoi.edu/institutes/coi/viewTopic.do?o=read&id=369; sewage total converted to liters.

6. Nellemann and Corcoran, op. cit. note 1, p. 16.

7. Kenneth R. Weiss, "Slowing a Tide of Pollutants," *Los Angeles Times*, 25 December 2006.

8. Don Hinrichsen, *Coastal Waters of the World: Trends, Threats and Strategies* (Washington, DC: Island Press, 1999), pp. 1, 7.

9. Ibid.

10. Nellemann and Corcoran, op. cit. note 1, pp. 6, 34.

11. Beman, Arrigo, and Matson, op. cit. note 1.

12. "U.N. Reports Growing Number of Ocean 'Dead Zones'," *Associated Press*, 19 October 2006.

13. Ibid.

14. Beman, Arrigo, and Matson, op. cit. note 1.

15. Richard C. Thompson et al., "Lost at Sea: Where Is All the Plastic?" *Science*, 7 May 2004, p. 838.

16. James Owen, "Oceans Awash With Microscopic Plastic, Scientists Say," *National Geographic News*, 6 May 2004.

17. Steve Connor, "Warmer Seas Will Wipe Out Plankton, Source of Ocean Life," (London) *The Independent*, 19 January 2006.

18. Scott C. Doney, "The Dangers of Ocean Acidification," *Scientific American*, May 2006; Elizabeth Kolbert, "The Darkening Sea," *The New Yorker*, 20 November 2006.

19. Kenneth R. Weiss, "Dark Tides, Ill Winds," *Los Angeles Times*, 1 August 2006; Kenneth R. Weiss, "Sentinels Under Attack," *Los Angeles Times*, 31 July 2006.

20. Oceana, *Needless Cruise Pollution: Passengers Want Sewage Dumping Stopped* (Washington, DC: 2003).

21. Ibid., data converted to liters.

22. Oceana, "Stop Cruise Ship Pollution," at www.oceana.org.

23. Oceana, "Royal Caribbean Campaign Victory," at www.oceana.org, 4 May 2004.

24. Nellemann and Corcoran, op. cit. note 1, p. 5.

25. Ibid., pp. 5, 15.

26. The Ocean Conservancy, "The Ocean Conservancy Announces Results of the International Coastal Cleanup; Next Cleanup Set for September," press release (Washington, DC: 20 July 2006).

27. The Ocean Conservancy, "The International Coastal Cleanup: A Legacy of Cleaner Oceans" (Washington, DC: 2006).

28. Ibid.

BOTTLED WATER CONSUMPTION JUMPS
(pages 102–03)

1. International Bottled Water Association, "The 2005 Stats: Bottled Water Continues as Number 2 in 2005" and "The 2002 Stats," Alexandria, VA, undated.

2. Beverage Marketing Corporation, *The Global Bottled Water Market, 2006 Edition* (New York: December 2006).

3. International Bottled Water Association, op. cit. note 1.

4. Ibid.

5. Ibid.

6. Ibid.

7. Peter H. Gleick, *The World's Water 2004–2005* (Washington, DC: Island Press, 2004), p. 25.

8. Catherine Ferrier, *Bottled Water: Understanding a Social Phenomenon*, Commissioned by World Wide Fund for Nature (Gland, Switzerland: April 2001), p. 17.

9. Beverage Marketing Corporation, *Selected Beverage Category Topline Report, 2006 Edition* (New York: May 2006).

10. Ferrier, op. cit. note 8, p. 18.

11. Ibid., p. 10.

12. Ibid., p. 19.

13. Gleick, op. cit. note 7, p. 23.

14. International Bottled Water Association, op. cit. note 1.

15. Gleick, op. cit. note 7, p. 40.

16. Uday Lal Pai, "Water—India Needs Massive Investment," for InvestorIdeas.com, 7 August 2006.

17. U.N. Development Programme, *Human Development Report 2006* (New York: Palgrave Macmillan, 2006), p. 36.

18. Gleick, op. cit. note 7, p. 26; Ferrier, op. cit. note 8, p. 20.

19. Gleick, op. cit. note 7, p. 26.

20. See www.codexalimentarius.net/web/index_en.jsp.

21. Natural Resources Defense Council, *Bottled Water: Pure Drink or Pure Hype* (Washington, DC: 1999).

22. Ibid.

23. Ibid.

24. Peter H. Gleick, *The World's Water 2006–2007* (Washington, DC: Island Press, 2006), p. 171.

25. Gleick, op. cit. note 7, p. 41.

26. Ferrier, op. cit. note 8, p. 18.

27. Ibid., p. 10.

28. Ibid., pp. 21, 22.

29. Brian C. Howard, "Despite the Hype, Bottled Water is Neither Cleaner nor Greener Than Tap Water," *E/The Environmental Magazine*, December 2003.

30. Container Recycling Institute, "Plastic Bottle Recycling Fails to Keep Up With Increasing Sales," press release (Washington, DC: 8 December 2006).

31. National Association for PET Container Resources, *2005 Report on Post Consumer PET Container Recycling Activity* (Sonoma, CA: undated), p. 2.

32. Ibid., p. 4.

33. Number of plastic water bottles provided by Patricia Franklin, executive director, Container Recycling Institute, Washington, DC, e-mail to author, 16 March 2007.

34. Container Recycling Institute, op. cit. note 30.

SUSTAINABLE COMMUNITIES BECOME MORE POPULAR (pages 104–05)

1. Quoted in Robert Gilman, "The Eco-village Challenge," *In Context*, summer 1991, p. 10.

2. Global Ecovillage Network (GEN), at gen.ecovillage .org/index.html, viewed 4 April 2007. Note: The number of ecovillages may include some that are still under construction.

3. GEN, op. cit. note 2.

4. Jonathan Dawson, *Ecovillages: New Frontiers for Sustainability* (Bristol, U.K.: The Schumacher Society, 2006), p. 21.

5. Author field visit to Earthaven Ecovillage and discussion with Diana Leafe Christian, editor of *Communities Magazine* and Earthaven resident, 14 March 2007.

6. Ibid.

7. Ibid.

8. Dawson, op. cit. note 4, pp. 26–28.

9. Ibid.; "Ecovillage at Mbam," *Global Ecovillage Network–Senegal*, at cresp.sn/gensen/defaulteng.htm, viewed 15 March 2007.

10. GEN, op. cit. note 2.

11. "Munksoegaard," *Intentional Community Database*, at icdb.org, viewed 4 April 2007.

12. GEN, op. cit. note 2.

13. Largest from ibid.; prize from Diana Leafe Christian, "Ecovillages: Where Are They, What Are They Doing?" *Earthlights Magazine*, May 2005.

14. Dawson, op. cit. note 4, pp. 32–34.

15. Ibid.

16. Ibid.

17. Ibid., p. 44.

18. Ibid., p. 29.

19. Eric Brende, *Better Off: Flipping the Switch on Technology* (New York: Harper Perennial, 2004).

20. Diana Leafe Christian, *Finding Community: How to Join an Ecovillage or Intentional Community* (Gabriola Island, BC: New Society Publishers, in press).

21. Miller McPherson, Lynn Smith-Lovin, and Matthew E. Brashears, "Social Isolation in America: Changes in Core Discussion Networks Over Two Decades," *American Sociological Review*, June 2006,

pp. 353–75.

22. Graham Meltzer, *Sustainable Community: Learning from the Cohousing Model* (Victoria, BC: Trafford, 2005), pp. 3–6.

23. Ibid.

24. Gary Gardner and Erik Assadourian, "Rethinking the Good Life," in Worldwatch Institute, *State of the World 2004* (New York: W. W. Norton & Company, 2004), pp. 171–72.

25. The Cohousing Association of the United States, "US Cohousing Communities," at directory.cohousing.org/us_list/all_us.php, viewed 4 April 2007; Canadian Cohousing Network, "Community Summary," at cohousing.ca/summary.htm, viewed 4 April 2007; Greg Bamford, "'Living Together on One's Own': Cohousing for Older People, A New Housing Type in Denmark and the Netherlands," conference paper, 2004; UK Co-housing Network, at www.cohousing.org.uk, viewed 4 April 2007; Sweden from Dick Urban Vestbro, Chair of National Association of Collective Housing Units, Kollektivhus NU, e-mail to author, 9 March 2007. Note: the number of cohousing communities includes communities that are still under construction or being planned.

26. Peabody Trust, *Peabody Trust Annual Review: Moving Communities Forward* (London: 2006); Peabody Trust, "BedZED (Beddington Zero Energy Development)," at www.peabody.org.uk/pages/GetPage.aspx?id=179, viewed 4 April 2007.

27. Peabody Trust, "BedZED," op. cit. note 26.

28. Dawson, op. cit. note 4, p. 44.

29. Hilary French, "Sacred Mountain," *World Watch Magazine*, May/June 2004, pp. 18–25; Nancy Chege, Kenyan National Coordinator of COMPACT, discussion with author, March 2006.

30. Intergovernmental Panel on Climate Change, *Climate Change 2007: The Physical Science Basis—Summary for Policymakers* (Geneva: 2007); Millennium Ecosystem Assessment, *Ecosystems and Human Well-being: Synthesis* (Washington, DC: Island Press, 2005), p. 1.

31. See Michael H. Shuman, *The Small-Mart Revolution: How Local Businesses Are Beating the Global Competition* (San Francisco: Berrett-Koehler Publishers, Inc., 2006).

32. Dan Chiras and Dave Wann, *Superbia: 31 Ways to Create Sustainable Neighborhoods* (Gabriola Island, BC: New Society Publishers, 2003).

33. "The Relocalization Network: Local Communities, Global Connections," at www.relocalize.net, viewed 3 April 2007.

34. "Ecovillage Training Center at The Farm," at www.thefarm.org/etc/courses.html, viewed 3 April 2007.

35. "Sarvodaya," at www.sarvodaya.org, viewed 4 April 2007.

36. Gary Gardner, *Invoking the Spirit: Religion and Spirituality in the Quest for a Sustainable World*, Worldwatch Paper 164 (Washington, DC: Worldwatch Institute, 2003), pp. 39–42.

PROGRESS TOWARD THE MDGS IS MIXED (pages 108–09)

1. Helen Nyambura-Mwaura, "Interview—Poor Nations Can Still Meet Poverty Goals—U.N.," *Reuters*, 22 January 2007; United Nations, *The Millennium Development Goals Report 2006* (New York: 2006), p. 3.

2. United Nations, "World Leaders Adopt 'United Nations Millennium Declaration' at Conclusion of Extraordinary Three-Day Summit," press release (New York: 8 September 2000).

3. Box 1 based on mdgs.un.org/unsd/mdg/Host.aspx?Content=Indicators/OfficialList, viewed 29 January 2007.

4. Ibid.

5. U.N. Development Programme (UNDP), *Human Development Report 2006* (New York: Palgrave Macmillan, 2006), p. 269.

6. United Nations, op. cit. note 1, p. 4.

7. Ibid.

8. World Bank, *Global Monitoring Report 2006* (Washington, DC: 2006), p. 3.

9. Decline in proportion of undernourished people from U.N. Food and Agriculture Organization, *The State of Food Insecurity in the World 2006* (Rome: 2006), p. 8; 834 million undernourished people is a preliminary estimate for 2002–04, available at www.fao.org/es/ess/faostat/foodsecurity/Files/NumberUndernourishment_en.xls, viewed 13 March 2007.

10. UNDP, op. cit. note 5, p. 267.

11. Ibid.

12. World Bank, op. cit. note 8, p. 4.

13. Ibid.

14. Ibid., pp. 4–5.

15. United Nations, op. cit. note 1, p. 15.

16. World Health Organization (WHO) and UNICEF, Joint Monitoring Programme (JMP) for Water Supply & Sanitation, at www.wssinfo.org/en/22_wat_global.html, viewed 12 March 2007. For definitions, see www.wssinfo.org/en/122_definitions.html, viewed 12 March 2007.

17. WHO and UNICEF, op. cit. note 16; WHO and UNICEF, JMP for Water Supply & Sanitation, at www.wssinfo.org/en/142_currentSit.html, viewed 12 March 2007.

18. WHO and UNICEF, JMP for Water Supply & Sanitation, at www.wssinfo.org/en/32_san_global.html, viewed 12 March 2007.

19. United Nations, op. cit. note 1, p. 18.

20. United Nations, "Highlights of Commitments and Implementation Initiatives," Johannesburg Summit 2002, 3 September 2002.

21. United Nations, "Press Conference on Millennium Development Goals by Special Advisor," New York, 20 December 2006.

22. Ibid.

23. United Nations, op. cit. note 1, p. 22.

24. World Bank, op. cit. note 8, p. 7.

25. United Nations, op. cit. note 1, p. 22.

26. Ibid., pp. 22–23.

LITERACY IMPROVES WORLDWIDE (pages 110–11)

1. UNESCO, Education For All Global Monitoring Report 2007 (Paris: 2006), p. 236. In Table 1, data before 1990 are from UNESCO, Education For All Global Monitoring Report 2006 (Paris: 2005), pp. 165–66.

2. United Nations, Millennium Development Goals Indicators, at mdgs.un.org.

3. UNESCO, Monitoring Report 2007, op. cit. note 1, pp. 2, 13.

4. Ibid., pp. 190, 236.

5. Ibid., p. 236.

6. UNESCO Institute for Statistics, "Adult Literacy Rates and Illiterate Population by Region and Gender," at www.uis.unesco.org/TEMPLATE/html/Exceltables/education/Literacy_Regional_CurrentRel.xls, viewed 10 January 2007.

7. Ibid.

8. UNESCO, Monitoring Report 2007, op. cit. note 1, p. 60.

9. Ibid., p. 25.

10. Ibid.

11. J. Wirt et al., eds., The Condition of Education 2004 (Washington, DC: U.S. Department of Education, National Center for Education Statistics), p. 54.

12. Ibid., p. 52.

13. Gina Kolata, "A Surprising Secret to a Long Life: Stay in School," New York Times, 3 January 2007.

14. UNESCO, Monitoring Report 2007, op. cit. note 1, p. 13.

15. Ibid., pp. 58, 59.

16. UNESCO Institute for Statistics, op. cit. note 6.

17. Brec Cooke, School of Education, Teaching, and Health, American University, Washington, DC, e-mail to author, 27 March 2007.

18. UNESCO, Monitoring Report 2007, op. cit. note 1, p. 2.

19. Ibid.

20. World Bank, "Girls' Education," at www.worldbank.org, viewed 10 January 2007.

21. UNESCO, Monitoring Report 2007, op. cit. note 1, p. 33.

22. World Bank, op. cit. note 20.

23. Stephanie Hanes, "Oprah's Academy: Why Educating Girls Pays Off More," Christian Science Monitor, 5 January 2007.

24. World Bank, op. cit. note 20.

25. UNESCO, Monitoring Report 2007, op. cit. note 1, p. 71.

26. Ibid., p. 74.

27. Ibid., p. 69.

28. Ibid., pp. 68–74.

29. Ibid., pp. 13, 85–86.

30. U.N. Development Programme (UNDP), Human Development Report 2003 (New York: Oxford University Press, 2003), pp. 266–69.

31. Ibid.

32. UNESCO, Monitoring Report 2007, op. cit. note 1, p. 70.

33. UNESCO, World Education Forum Final Report (Paris: 2000), pp. 43–69.

34. James Heckman, "Catch 'Em Young," Wall Street Journal, 10 January 2006.

35. UNESCO, Monitoring Report 2007, op. cit. note 1, p. 113.

36. Ibid., p. 133.

37. Ibid., p. 22.

38. UNDP, op. cit. note 30, p. 87.

39. Ibid.

40. UNESCO, *Monitoring Report 2007*, op. cit. note 1, pp. 154–61.

CHILD LABOR HARMS MANY YOUNG LIVES
(pages 112–13)

1. International Labour Organization (ILO), *The End of Child Labour: Within Reach* (Geneva: 2006), p. 6. The ILO says that "the concept of 'child labor' is based on the ILO Minimum Age Convention, 1973 (No. 138)."

2. Ibid.

3. Ibid.

4. Ibid.

5. Ibid.

6. Ibid. Economic activity, as defined here, includes "most productive activities undertaken by children," whether paid or not, full time or not, legal or illegal. It excludes schooling and chores done in a child's own household.

7. Frank Hagemann et al., *Global Child Labour Trends: 2000 to 2004* (Geneva: International Program on the Elimination of Child Labor, ILO, 2006), p. 3. Data are for economically active children 5–14 years old.

8. Ibid.

9. Ibid.

10. Ibid.

11. ILO, "Facts on Child Labour," no date, at www.ilo .org/public/english/bureau/inf/download/child/ childday06.pdf.

12. Ibid.

13. ILO, op. cit. note 1, p. 14; "Global Child Labour Figures Fall," *BBC News*, 4 May 2006.

14. ILO, op. cit. note 1, p. 14; Progressive Policy Institute, "Child Labor Rates are Falling," Trade Fact of the Week, 10 May 2006.

15. Roxanne Lawson and Tim Newman, *Stopping Firestone: Getting Rubber to Meet the Road* (Silver City, NM, and Washington, DC: Foreign Policy In Focus, December 2006).

16. Ibid.

17. Stop Firestone Campaign, at www.stopfirestone.org.

18. *John Doe I et al. v. Nestlé et al.*, "Class Action Complaint for Injunctive Relief and Damages," Case CV-05-5133-SVW, p. 9.

19. "Child-labor Chocolates" (editorial), *Los Angeles Times*, 14 February 2007.

20. Tex Dworkin, "The Bitter Truth About Chocolate," *Treehugger.com*, 1 February 2007.

21. Ibid.

22. Ibid.

23. *John Doe I et al. v. Nestlé et al.*, "Certain Defts.' Opening Brief in Response to Court's July 27, 2006 Order for Further Briefing," 9 August 2006, Case CV-05-5133-SVW.

24. Timothy Newman, International Labor Rights Fund, e-mail to author, 27 February 2007.

25. U.S. Department of Labor, "US Labor Department Funds Project to Evaluate Effectiveness of Anti-Child-Labor Efforts in the Cocoa Industry," press release (Washington, DC: 3 October 2006).

26. Ibid.

27. ILO, op. cit. note 1, p. 8.

28. Ibid.

29. "Child Labor," Human Rights Watch, at hrw.org/ children/labor.htm, viewed 7 February 2007.

30. Ibid.

31. Ibid.

32. Ibid.

33. Sharon LaFraniere, "Africa's World of Forced Labor, in a 6-Year-Old's Eyes," *New York Times*, 20 October 2006.

34. Ibid.

35. Bureau of International Labor Affairs, *2005 Findings on the Worst Forms of Child Labor* (Washington, DC: U.S. Department of Labor, 2006), p. xxii.

36. LaFraniere, op. cit. note 33.

37. Ibid.

38. Child Labor Act from Renu Agal, "Misery Goes on for India Child Workers," *BBC News*, 10 December 2006; bonded laborers from "Child Labor," op. cit. note 29.

39. Agal, op. cit. note 38.

40. Ibid.

41. "Child Labor," op. cit. note 29.

42. Ibid.

INFORMAL ECONOMY THRIVES IN CITIES
(pages 114–15)

1. Martha Alter Chen, "Rethinking the Informal Economy: Linkages with the Formal Economy and the Formal Regulatory Environment," *EGDI and UNU-WIDER*, April 2005, p. 11.

2. Ibid., p. 13; "The Global Workforce: A Statistical Picture," Women in Informal Employment: Globalizing and Organizing, at wiego.org/stat_picture.

3. "La Informalidad No Cede Terreno y Cobija a más de la Mitad de los Trabajadores Colombianos," *El Tiempo*, 19 February 2007.

4. Norman V. Loayza, *The Economics of the Informal Sector: A Simple Model and Some Empirical Evidence from Latin America* (Washington, DC: World Bank, 1997), p. 47.

5. Hernando de Soto, *The Mystery of Capital: Why Capitalism Triumphs in the West and Fails Everywhere Else* (New York: Basic Books, 2000), p. 30.

6. Vincent Palmade and Andrea Anayiotos, "Rising Informality," *Public Policy for the Private Sector*, Note No. 298, August 2005.

7. Friedrich Schneider, *The Size of the Shadow Economies of 145 Countries All Over the World: First Results Over the Period 1999–2003* (Bonn, Germany: Institute for the Study of Labor, December 2004).

8. Palmade and Anayiotos, op. cit. note 6.

9. Kai N. Lee, "An Urbanizing World," in Worldwatch Institute, *State of The World 2007* (New York: W. W. Norton & Company, 2007), p. 5.

10. Central Intelligence Agency, *The World Factbook: India*, at www.cia.gov/cia/publications/factbook/print/in.html.

11. Diego Palma, *La Informalidad, Lo Popular y el Cambio Social* (Lima, Peru: Centro de Estudios y Promoción del Desarrollo, 1987), pp. 17–18; de Soto, op. cit. note 5.

12. Hernando de Soto and Francis Cheneval, eds., *Realizing Property Rights* (Italy: Buffer&Rub, 2006), p. 21.

13. Ibid., p. 26.

14. Ibid., p. 51.

15. Carlos Ramon Ponze Monteza, *Gamarra: Formación Estructura y Perspectivas* (Lima, Peru: Fundación Friedrich Ebert, 1994), p. 66.

16. Palma, op. cit. note 11, p. 22.

17. Fernando Escalante Gonzalbo, "La Nueva Política," *Cronica*, 27 April 2005.

18. Lee, op. cit. note 9, p. 5.

19. Chen, op. cit. note 1, p. 19.

20. John Lancaster, "Next Stop, Squalor," *Smithsonian*, March 2007.

21. Ibid.

22. Ibid.

23. De Soto, op. cit. note 5, p. 33.

24. Chen, op. cit. note 1, p. 11.

25. Debraj Ray, *Development Economics* (Princeton, NJ: Princeton University Press, 1998), p. 347.

26. Lancaster, op. cit. note 20.

27. Ibid.

28. De Soto, op. cit. note 5, p. 49.

29. Karol Boudreaux, "Property Rights and Resource Conflict in the Sudan," in de Soto and Cheneval, op. cit. note 12, p . 72; Juan Pablo Viqueira, *Encrucijadas Chiapanecas* (Mexico: Tusquets/COLMEX, 2002).

30. Celia W. Dugger, "In Bangalore, India, a Cuddle with Your Baby Requires a Bribe," *New York Times*, 30 August 2005.

31. Ibid.

32. De Soto, op. cit. note 5, p. 32.

33. Alexandra C. Horst, *2007 International Property Rights Index* (Washington, DC: Property Rights Alliance, 2007), p. 32.

SOCIALLY RESPONSIBLE INVESTMENT GROWS RAPIDLY (pages 116–17)

1. Social Investment Forum, *2005 Report on Socially Responsible Investing Trends in the United States: 10-Year Review* (Washington, DC: 24 January 2006), p. v. Totals in Table 1 were converted to U.S. dollars using historical exchange rate values, from *FX History: Historical Currency Exchange Rates*, at www.oanda.com/convert/fxhistory, viewed 9 March 2007.

2. Eurosif, *European SRI Study 2006* (Paris: 2006), p. 5.

3. Social Investment Organization, *Canadian Social Investment Review 2007* (Toronto, ON: March 2007), p. 5; Corporatemonitor, *Sustainable Responsible Investment in Australia–2006* (Evans Head, Australia: September 2006), p. 4; Sho Ikeda, managing director, Cangen Biotechnologies, Japan, e-mail to author, 14 February 2007.

4. Owens, Williams, & Wood Consulting, "CSR Index for Malaysia Launched," *CSRwire*, 5 December 2006.

5. Ikeda, op. cit. note 3; Social Investment Forum, op. cit. note 1, p. iv.

6. Corporatemonitor, op. cit. note 3, p. 4.

7. Social Investment Forum, op. cit. note 1, p. v.

8. Social Investment Organization, op. cit. note 3, p. 5.

9. Eurosif, op. cit. note 2, p. 11.

10. Jonathan Schuman, vice president for product and business development, AIG Global Investment

Notes

Group, Japan, e-mail to author, 19 February 2007.

11. Social Investment Forum, op. cit. note 1, p. v.

12. Ibid.

13. Ethical Funds, "Shareholder Resolutions 2006," at www.ethicalfunds.com/Advisor/sri/resolutions_2006 .asp, viewed 23 February 2007.

14. Ibid.

15. Patricia Wolf, Executive Director, Interfaith Center on Corporate Responsibility, discussion with author, 25 April 2005.

16. Social Investment Forum, op. cit. note 1, p. v.

17. KLD, "Long-term Performance of 15 Year-Old Domini 400 Social Index Helps Validate SRI," at www.kld.com/article.cgi?id=1703, viewed 12 February 2007.

18. Mark Campanale, member of the Advisory Board to 3iG, "New Investments for the Faiths: Opportunities or Sources of Conflict," PowerPoint presentation at TBLI Faith-Consistent Investment Seminar, Paris, 9 November 2006.

19. Bob Willard, *The NEXT Sustainability Wave* (Gabriola Island, BC: New Society Publishers, 2005), p. 135.

20. Bill Baue, "Global 100 List Raises the Profile of Sustainability at the World Economic Forum (and Beyond)," *SRI-adviser.com*, 25 January 2007.

21. Eurosif, op. cit. note 2, p. 12.

22. Ibid.

23. "Financial Giants Push Social Investing into the Mainstream," *SRIWorld*, 10 January 2001.

24. Social Investment Organization, op. cit. note 3, p. 5.

25. Eurosif, op. cit. note 2, p. 12.

26. Bill Baue, "Microfinance Crosses Continental Divide with $100 Million Commitment from TIAA-CREF," *SRI-adviser.com*, 9 October, 2006.

27. Bill Baue, "Top Five Socially Responsible Investing News Stories of 2006," *SRI-adviser.com*, 5 January 2007.

28. Ibid.

29. Ibid.

30. 3iG, "About Us," at www.3ignet.org/about/members .html, viewed 6 February 2007.

31. Ibid.

HIV/AIDS CONTINUES WORLDWIDE CLIMB
(pages 120–21)

1. Estimates based on Joint United Nations Programme on HIV/AIDS (UNAIDS) and World Health Organization (WHO), *2006 AIDS Epidemic Update* (Geneva: December 2006). UNAIDS's estimate of people living with HIV for 2006 is 34.1–47.1 million worldwide; estimates of new HIV infections range from 3.6 million to 6.6 million; estimates of deaths due to AIDS in 2006 range from 2.5 million to 3.5 million.

2. UNAIDS and WHO, op. cit. note 1.

3. Ibid.

4. Ibid.

5. Ibid.

6. Ibid.

7. Ibid.

8. Ibid.

9. Ibid.

10. Ibid.

11. Ibid.

12. Ibid.

13. Ibid.

14. Ibid.

15. Ibid.

16. Ibid., p. 5.

17. Joyce Mulamy, "End to HIV/AIDS: A Tall Order in Face of Violence," *World Social Forum*, 24 January 2007.

18. Ibid.

19. UNAIDS and WHO, *2006 Report on the Global AIDS Epidemic: Progress in Countries* (Geneva: 2006).

20. UNAIDS and WHO, op. cit. note 1.

21. O. Abdelwahab, "Prevalence, Knowledge of AIDS and HIV Risk-related Sexual Behavior among Police Personnel in Khartoum State, Sudan," XVI International AIDS Conference, Toronto, 13–18 August 2006.

22. UNAIDS and WHO, op. cit. note 1.

23. UNAIDS, *2006 Report on the Global AIDS Epidemic: Financing the Response to AIDS* (Geneva: 2006).

24. Ibid.

25. Ibid.

26. David Brown, "Gates Foundation Giving $500 Million to Fight Disease," *Washington Post*, 10 August 2006.

27. Andrew Jack, "The Drug Companies: A New Mood of Co-operation," *Financial Times*, 1 December 2006.

28. Accelerating Access Initiative, *The International Federation of Pharmaceutical Manufacturers & Associations*, at www.ifpma.org.

29. Ibid.
30. Jack, op. cit. note 27.
31. Ibid.
32. John Donnelly, "U.S. Buying More Generic AIDS Drugs," *Boston Globe*, 12 November 2006.
33. Ibid.
34. Indira A. R. Lakshmanan, "The Rising Cost of AIDS Drugs Threatens Brazil's Free Treatment Program," *International Herald Tribune*, 3 January 2007.
35. Ibid.
36. Ibid.

MALARIA REMAINS A THREAT (pages 122–23)

1. Roll Back Malaria (RBM), World Health Organization (WHO), and UNICEF, *World Malaria Report 2005* (Geneva: 2005), p. xvii.
2. World Economic Forum (WEF), *Business and Malaria: A Neglected Threat?* (Geneva: 2006), pp. 9–11.
3. Robert W. Snow et al., "The Global Distribution of Clinical Episodes of *Plasmodium falciparum* Malaria," *Nature*, 10 March 2005, pp. 214–17.
4. RBM, WHO, and UNICEF, op. cit. note 1, p. 5; Centers for Disease Control and Prevention, "Biology of Malaria," at www.cdc.gov/malaria/biology/index.htm, viewed 1 April 2007.
5. Darwin H. Stapleton, "Lessons of History? Antimalaria Strategies of the International Health Board and the Rockefeller Foundation from the 1920s to the Era of DDT," *Public Health Reports*, March–April 2004, pp. 211–12.
6. "Malaria History," at www.malariasite.com/malaria/history_literature.htm, viewed 1 April 2007.
7. Caterina Guinovart et al., "Malaria: Burden of Disease," *Current Molecular Medicine*, March 2006, p. 138.
8. RBM, WHO, and UNICEF, op. cit. note 1, p. xv.
9. Jeffrey Sachs and Pia Malaney, "The Economic and Social Burden of Malaria," *Nature*, 7 February 2002, pp. 680–85.
10. WEF, op. cit. note 2, p. 28.
11. Global Fund to Fight HIV/AIDS, Tuberculosis and Malaria, *HIV/AIDS, Tuberculosis and Malaria: The Status and Impact of the Three Diseases* (Geneva: 2004), p. 20.
12. Carlos A. Guerra, Robert W. Snow, and Simon I. Hay, "Mapping the Global Extent of Malaria in 2005," *Trends in Parasitology*, August 2006, pp. 353–58.
13. Ronald Ross, "The Role of the Mosquito in the Evolution of the Malarial Parasite: The Recent Researches of Surgeon-Major Ronald Ross, I.M.S. 1898," *Yale Journal of Biology and Medicine*, vol. 75, no. 2 (2002), pp. 103–05.
14. Stapleton, op. cit. note 5.
15. Peter J. Brown, "Failure-as-Success: Multiple Meanings of Eradication in the Rockefeller Foundation Sardinia Project, 1946–1951," *Parassitologia*, June 1998, pp. 117–30; Amir Attaran et al., "Balancing Risks on the Backs of the Poor," *Nature Medicine*, July 2000, pp. 729–31.
16. Rachel L. Carson, *Silent Spring* (New York: Houghton Mifflin Company, 1962), p. 400; Attaran et al., op. cit. note 15; Tina Rosenberg, "What the World Needs Now is DDT," *New York Times*, 11 April 2004.
17. Amir Attaran and MFI International Board, "Open Letter to Stockholm Convention POPs Treaty Delegate Negotiators," 1999, at www.malaria.org/DDTpage.html, viewed 1 April 2007.
18. U.N. Environment Programme, *Stockholm Convention on Persistent Organic Pollutants (POPs)*, at www.pops.int, Annex B.
19. Apoorva Mandavilli, "Health Agency Backs Use of DDT Against Malaria," *Nature*, 21 September 2006, pp. 250–51; U.S. Agency for International Development, "Indoor Residual Spraying (IRS)," 2006, at www.usaid.gov/our_work/global_health/id/malaria/techareas/irs.html, viewed 1 April 2007; Roger Bate, Philip Coticelli, and Richard Tren, *Moving Mountains: The Evolution of USAID's Malaria Control Program* (Washington, DC: American Enterprise Institute, 2006), p. 2.
20. Nicholas J. White, "Malaria—Time to Act," *New England Journal of Medicine*, 9 November 2006, pp. 1956–57; "Why the World Needs Another Malaria Initiative," *The Lancet*, 31 July 2004, pp. 389–90.
21. White, op. cit. note 20; "Why the World Needs Another Malaria Initiative," op. cit. note 20.
22. White, op. cit. note 20; Peter G. Kremsner and Sanjeev Krishna, "Antimalarial Combinations," *The Lancet*, 17 July 2004, pp. 285–94.
23. Amir Attaran et al., "WHO, the Global Fund, and Medical Malpractice in Malaria Treatment," *The Lancet*, 17 January 2004, pp. 237–40.
24. Declan Butler, "Global Fund Changes Tack on

Notes

Malaria Therapy," *Nature*, 10 June 2004, p. 588.

25. Christoph Benn and Bernhard Schwartlander, *The Resource Needs of the Global Fund 2005–2007* (Geneva: Global Fund to Fight AIDS, Tuberculosis and Malaria, 2005), pp. 16–18.

26. WHO, *Meeting on the Production of Artemisinin and Artemisinin Based Combination Therapies* (Geneva: 2005).

27. RBM, WHO, and UNICEF, op. cit. note 1, p. 17.

28. Maria E. Rafel et al., "Reducing the Burden of Childhood Malaria in Africa: The Role of Improved Diagnosis," *Nature*, 27 November 2006, pp. 39–48.

29. "Finally Clearing the Air," *The Economist*, 7 December 2006, pp. 63–64.

30. Christian Lengeler and Brian Sharp, "Indoor Residual Spraying and Insecticide-Treated Nets," in Global Health Council, *Reducing Malaria's Burden. Evidence of Effectiveness for Decision Makers* (White River Junction, VT: 2003), pp. 21–22.

31. Sarah Boseley, "Arata Kochi: Shaking Up the Malaria World," *The Lancet*, 17 June 2006, p. 1973.

32. Malaria Foundation International, "Malaria Advocacy—The Beginnings," at www.malaria.org, viewed 1 April 2007.

33. Joel G. Breman, Martin S. Alilio, and Anne Mills, "Conquering the Intolerable Burden of Malaria: What's New, What's Needed: A Summary," *American Journal of Tropical Medicine and Hygiene*, August 2004, pp. 1–15; Sylvia Meek et al., *Tackle Malaria Today: Give Tomorrow a Chance* (London: Medicines for Malaria Venture, 2005).

34. Sachs and Malaney, op. cit. note 9; Media Tenor, "Malaria Not on the Media Agenda," at www.medi atenor.com/newsletters.php?id_news=171, viewed 1 April 2007.

35. Bill & Melinda Gates Foundation, "Major Commitment to Global Fight Against Malaria," at www.gates foundation.org/GlobalHealth/Pri_Diseases/Malaria/ Announcements/Announce-061211.htm, viewed 1 April 2007; Jean-Louis Sarbib, Gobind Nankani, and Praful Patel, "The Booster Program for Malaria Control: Putting Knowledge and Money to Work," *The Lancet*, 15 July 2006, pp. 253–57; President's Malaria Initiative, "Fast Facts: The President's Malaria Initiative (PMI)," at www.fightingmalaria.gov/ resources/pmi_fastfacts.pdf, viewed 1 April 2007.

36. Peter. M. Nyarango et al., "A Steep Decline of Malaria Morbidity and Mortality Trends in Eritrea

between 2000 and 2004: The Effect of Combination of Control Methods," *Malaria Journal*, 24 April 2006, p. 33.

37. Chantal M. Morel, Jeremy A. Lauer, and David B. Evans, "Cost Effectiveness Analysis of Strategies to Combat Malaria in Developing Countries," *British Medical Journal*, 3 December 2005, p. 1299.

38. Malaria R&D Alliance, *Malaria Research & Development: An Assessment of Global Investment* (Seattle, WA: Program for Appropriate Technology in Health, 2005), p. 7.

39. Ibid.

40. Ibid.; Malaria R&D Alliance, "Innovation and the Defeat of Malaria," at www.mmv.org/IMG/pdf/mra _3.pdf.

41. Marc P. Girard et al., "A Review of Human Vaccine Research and Development: Malaria," *Vaccine*, 19 February 2007, pp. 1567–80.

42. Malaria R&D Alliance, op. cit. note 37; Malaria R&D Alliance, op. cit. note 39.

MALE REPRODUCTIVE HEALTH DECLINES (pages 124–25)

1. Testicular cancer rates from World Health Organization (WHO), International Agency for Research on Cancer (IARC), *IARC CancerBase* (Lyon, France: updated 2005); sperm count from Shanna H. Swan, Eric P. Elkin, and Laura Fenster, "The Question of Declining Sperm Density Revisited: An Analysis of 101 Studies Published 1934–1996," *Environmental Health Perspectives*, October 2000, pp. 961–66.

2. E. Huyghe, T. Matsuda, and P. Thonneau, "Increasing Incidence of Testicular Cancer Worldwide: A Review," *Journal of Urology*, July 2003, pp. 5–11.

3. J. Toppari et al., "Male Reproductive Health and Environmental Xenoestrogens," *Environmental Health Perspectives*, August 1996, pp. 741–803.

4. N. E. Skakkebaek et al., "Testicular Dysgenesis Syndrome: An Increasingly Common Developmental Disorder with Environmental Aspects," *Human Reproduction*, May 2001, pp. 972–78.

5. N. E. Skakkebaek, "Testicular Dysgenesis Syndrome: New Epidemiological Evidence," *International Journal of Andrology*, August 2004, pp. 189–91.

6. Swan, Elkin, and Fenster, op. cit. note 1.

7. H. Fisch et al., "Semen Analyses in 1,283 Men from the United States over a 25-Year Period: No Decline

in Quality," *Fertility and Sterility*, May 1996, pp. 1009–14.

8. C. A. Paulsen et al., "Data from Men in Greater Seattle Area Reveals No Downward Trend in Semen Quality: Further Evidence that Deterioration of Semen Quality is Not Geographically Uniform," *Fertility and Sterility*, May 1996, pp. 1015–20; Fisch et al., op. cit. note 7.

9. Shanna H. Swan et al., "Geographic Differences in Semen Quality of Fertile U.S. Males," *Environmental Health Perspectives*, April 2003, pp. 414–20; Niels Jørgensen et al., "European Differences from Regional Differences in Semen Quality in Europe," *Human Reproduction*, May 2001, pp. 1012–19.

10. J. P. E. Bonde et al., "Relation Between Semen Quality and Fertility: A Population-based Study of 430 First-pregnancy Planners," *Lancet*, April 1999, pp. 1172–77.

11. N. Jorgensen et al., "Coordinated European Investigations of Semen Quality: Results from Studies of Scandinavian Young Men is a Matter of Concern," *International Journal of Andrology*, February 2006, pp. 54–61, discussion pp. 105–08.

12. E. Huyghe, T. Matsuda, and P. Thonneau, "Increasing Incidence of Testicular Cancer Worldwide: A Review," *Journal of Urology*, July 2003, pp. 5–11.

13. Ibid.

14. WHO, op. cit. note 1.

15. L. A. G. Ries et al., eds., *SEER Cancer Statistics Review, 1973–1995* (Bethesda, MD: National Cancer Institute. 1998).

16. J. Richthoff et al., "Higher Sperm Counts in Southern Sweden Compared with Denmark," *Human Reproduction*, September 2002, pp. 2468–73.

17. D. J. Cahill and P. G. Wardle, "Clinical Review, Management of Infertility," *BMJ*, 6 July 2002, pp. 28–32.

18. Ibid.

19. A. Agarwal et al., "Relationship between Cell Phone Use and Human Fertility: An Observational Study," *Highlights from the 62nd Annual Meeting of the American Society for Reproductive Medicine*, October 2006, p. 398.

20. Ibid.

21. A. O. Cheek and J. A. McLachlan, "Environmental Hormones and the Male Reproductive System," *Journal of Andrology*, January/February 1998, pp. 5–10.

22. Theo Colborn et al., *Our Stolen Future* (New York: Penguin, 1996), p. 81; Alison Abbott "…While Japan Studies Drop in Sperm Counts," *Nature*, October 1998, p. 828.

23. "Prague Declaration on Endocrine Disruption," 10–12 May 2005, updated 26 May 2006.

24. S. M. Duty et al., "Phthalate Exposure and Human Semen Parameters," *Epidemiology*, May 2003, pp. 269–77.

25. Guowei Pan et al., "Decreased Serum Free Testosterone in Workers Exposed to High Levels of Di-n-butyl Phthalate (DBP) and Di-2-ethylhexyl Phthalate (DEHP): A Cross-Sectional Study in China," *Environmental Health Perspectives*, November 2006, pp. 1643–47.

26. Duty et al., op. cit. note 24.

27. R. H. Kaufman et al., "Continued Follow-up of Pregnancy Outcomes in Diethylstilbestrol-Exposed Offspring," *Obstetrics & Gynecology*, October 2000, pp. 483–89.

28. T. K. Jensen et al., "Do Environmental Estrogens Contribute to the Decline in Male Reproductive Health?" *Clinical Chemistry*, December 1995, pp. 1896–1901.

29. U.S. Environmental Protection Agency, "Endocrine Disruptor Screening Program," at www.epa.gov/scipoly/oscpendo/pubs/edspoverview/primer.htm, updated March 2006.

30. Glen Hess, "Endocrine Distruption Screening Will Evaluate 15,000 Chemicals," *Chemical Market Reporter*, October 1998.

31. "Prague Declaration on Endocrine Disruption," op. cit. note 23.

32. T. B. Hayes et al., "Pesticide Mixtures, Endocrine Disruption, and Amphibian Declines: Are We Underestimating the Impact?" *Environmental Health Perspectives*, supp. 1, April 2006, pp. 40–50.

33. "Prague Declaration on Endocrine Disruption," op. cit. note 23.

The Vital Signs Series

Some topics are included each year in *Vital Signs*; others are covered only in certain years. The following is a list of topics covered in *Vital Signs* thus far, with the year or years they appeared indicated in parentheses. Those marked with a bullet (•) appeared in Part One, which includes time series of data on each topic; 2006 indicates *Vital Signs 2006–2007*; 2007 indicates this edition of *Vital Signs*.

AGRICULTURE AND FOOD

Agricultural Resources

- Fertilizer Use (1992–2001)
- Grain Area (1992–93, 1996–97, 1999–2000)
- Grain Yield (1994–95, 1998)
- Irrigation (1992, 1994, 1996–99, 2002, 2007)

  Livestock (2001)

  Organic Agriculture (1996, 2000)

  Pesticide Control or Trade (1996, •2000, 2002, •2006)

  Transgenic Crops (1999–2000)

  Urban Agriculture (1997)

Food Trends

- Aquaculture (1994, 1996, 1998, 2002, 2005)

  Biotech Crops (2001–02)
- Cocoa Production (2002)
- Coffee (2001)

  Eggs (2007)
- Fish (1992–2000, 2006–07)
- Grain Production (1992–2003, 2005–07)
- Grain Stocks (1992–99)

- Grain Used for Feed (1993, 1995–96)
- Meat (1992–2000, 2003, 2005–07)
- Milk (2001)
- Soybeans (1992–2001, 2007)
- Sugar and Sweetener Use (2002)

THE ECONOMY

Resource Economics

  Agricultural Subsidies (2003)
- Aluminum (2001, 2006–07)

  Arms and Grain Trade (1992)

  Commodity Prices (2001)

  Fossil Fuel Subsidies (1998)
- Gold (1994, 2000, 2007)

  Illegal Drugs (2003)

  Metals Exploration (1998, •2002)
- Metals Production (2002)
- Paper (1993, 1994, 1998–2000)

  Paper Recycling (1994, 1998, 2000)
- Roundwood (1994, 1997, 1999, 2002, 2006–07)

  Seafood Prices (1993)
- Steel (1993, 1996, 2005–07)

  Steel Recycling (1992, 1995)

Farmland Quality (2002)
Forests (1992, 1994–98, 2002, 2005–06)
Groundwater (2000, 2006)
Ice Melting (2000, 2005)
Mangroves (2006)
Plant Diversity (2006)
Terrestrial Biodiversity (2007)
Water Scarcity (1993, 2001–02)
Water Tables (1995, 2000)
Wetlands (2001, 2005)

Natural Resource Uses

Biomass Energy (1999)
Dams (1995)
Ecosystem Conversion (1997)
Energy Productivity (1994)
Organic Waste Reuse (1998)
Soil Erosion (1992, 1995)
Tree Plantations (1998)

Pollution

Acid Rain (1998)
Air Pollution (1993, 1999, 2005)
Algal Blooms (1999)
Hazardous Wastes (2002)
Lead in Gasoline (1995)
Mercury (2006)
Nuclear Waste (1992, •1995)
Ocean (2007)
Pesticide Resistance (•1994, 1999)
• Sulfur and Nitrogen Emissions (1994–97)

Other Environmental Topics

Bottled Water (2007)
Environmental Indicators (2006)
Environmental Treaties (•1995, 1996, 2000, 2002)
Invasive Species (2007)
Nitrogen Fixation (1998)
Pollution Control Markets (1998)
Sea Level Rise (2003)
Semiconductor Impacts (2002)
Transboundary Parks (2002)
• World Heritage Sites (2003)

THE MILITARY

• Armed Forces (1997)
Arms Production (1997)
• Arms Trade (1994)
Landmines (1996, 2002)
• Military Expenditures (1992, 1998, 2003, 2005–06)
• Nuclear Arsenal (1992–96, 1999, 2001, 2005, 2007)
Peacekeeping Expenditures (1993, •1994–2003, •2005–07)
Resource Wars (2003)
• Wars (1995, 1998–2003, 2005–07)
Small Arms (1998–99)

SOCIETY AND HUMAN WELL-BEING

Health

AIDS/HIV Incidence (•1994–2003, •2005–06, 2007)
Alternative Medicine (2003)
Asthma (2002)
Avian Flu (2007)
Breast and Prostate Cancer (1995)
• Child Mortality (1993)
• Cigarettes (1992–2001, 2003, 2005)
Drug Resistance (2001)
Endocrine Disrupters (2000)
Food Safety (2002)
Hunger (1995)
• Immunizations (1994)
• Infant Mortality (1992, 2006)
Infectious Diseases (1996)
Life Expectancy (1994, •1999)
Malaria (2001, 2007)
Malnutrition (1999)
Mental Health (2002)
Mortality Causes (2003)
Noncommunicable Diseases (1997)
Obesity (2001, 2006)
• Polio (1999)
Sanitation (1998)
Soda Consumption (2002)

Traffic Accidents (1994)
Tuberculosis (2000)
Water and Sanitation (1995, 2006)

Reproduction and Women's Status

Family Planning Access (1992)
Female Education (1998)
Fertility Rates (1993)
Maternal Mortality (1992, 1997, 2003)
 • Population Growth (1992–2003,
  2005–07)
Sperm Count (1999, 2007)
Violence Against Women (1996, 2002)
Women in Politics (1995, 2000)

Social Inequities

Homelessness (1995)
Income Distribution (1992, 1995, 1997,
 2002–03)
Language Extinction (1997, 2001, 2006)
Literacy (1993, 2001, 2007)
Prison Populations (2000)
Slums (2006)
Social Security (2001)
Teacher Supply (2002)
Unemployment (1999, 2005)

Other Social Topics

Aging Populations (1997)
Child Labor (2007)
Fast-Food Use (1999)
International Criminal Court (2003)
Millennium Development Goals (2005,
 2007)
Nongovernmental Organizations (1999)
Orphans Due to AIDS Deaths (2003)
Public Policy Networks (2005)
Quality of Life (2006)
Refugees (• 1993–2000, 2001, 2003,
 • 2005)
Religious Environmentalism (2001)
Sustainable Communities (2007)
Urbanization (• 1995–96, • 1998, • 2000,
 2002, • 2007)

Voter Turnouts (1996, 2002)
Wind Energy Jobs (2000)

TRANSPORTATION AND COMMUNICATIONS

 • Air Travel (1993, 1999, 2005–07)
 • Automobiles (1992–2003, 2005–07)
 • Bicycles (1992–2003, 2005–07)
  Car-sharing (2002, 2006)
  Computer Production and Use (1995)
  Gas Prices (2001)
  Electric Cars (1997)
 • Internet (1998–2000, 2002)
 • Internet and Telephones Combined
  (2003, 2006–07)
 • Motorbikes (1998)
 • Railroads (2002)
 • Satellites (1998–99)
 • Telephones (1998–2000, 2002)
  Urban Transportation (1999, 2001)

VITAL SIGNS ONLINE

Over 100 global indicators are now available online at...

www.worldwatch.org/vsonline

Categories of trends:
- Food and Agriculture
- Energy and Climate
- Global Economy and Resources
- Transportation and Communications
- Population and Society
- Health and Disease
- Conflict and Peace
- Environment

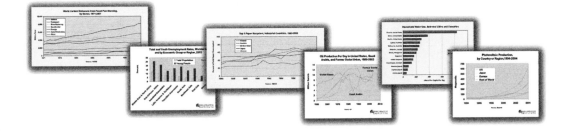

For each trend you can download the analysis and the data, charts, and graphs both in Excel spreadsheet format and in PowerPoint so you can add them to your own reports and presentations about the trends shaping our future. Get one trend or several from the eight different categories.

2 0 0 7
STATE OF THE WORLD
Our Urban Future

In 2008, half of the Earth's population will live in urban areas, marking the first time in history that humans are an urban species. *State of the World 2007: Our Urban Future* explores the myriad ways urbanization is affecting our lives and the global environment—with a special focus on the ideas that can make our cities environmentally sustainable and healthier places to live.

TABLE OF CONTENTS:

Chapter 1. **An Urbanizing World**

Chapter 2. **Providing Clean Water and Sanitation**

Chapter 3. **Farming the Cities**

Chapter 4. **Greening Urban Transportation**

Chapter 5. **Energizing Cities**

Chapter 6. **Reducing Natural Disaster Risk in Cities**

Chapter 7. **Charting a New Course for Urban Public Health**

Chapter 8. **Strengthening Local Economies**

Chapter 9. **Fighting Poverty and Environmental Injustice in Cities**

WITH CASE STUDIES OF THESE CITIES:

Timbuktu, Mali: Greening the Hinterlands

Loja, Ecuador: Ecological and Healthy City

Lagos, Nigeria: Collapsing Infrastructure

Freetown, Sierra Leone: Urban Farms After a War

Los Angeles, U.S.A.: End of Sprawl

Melbourne, Australia: Reducing a City's Carbon Emissions

Rizhao, China: Solar-Powered City

Malmö, Sweden: Building a Green Future

Jakarta, Indonesia: River Management

Mumbai, India: Policing by the People

Nairobi, Kenya: Life in Kibera

Petra, Jordan: Managing Tourism

Brno, Czech Republic: Brownfield Redevelopment

ORDER YOUR COPY TODAY!

Four Easy Ways To Order:

▶▶ Call us toll-free at 1-877-539-9946

▶▶ Fax us at 1-301-567-9553

▶▶ Visit us at www.worldwatch.org

▶▶ Mail your order to:

Worldwatch Institute

PO Box 879

Oxon Hill, MD 20750-0910

ISBN 978-0-393-32923-0
$18.95 plus S&H

For press and academic review copies, contact Julia Tier, jtier@worldwatch.org

Worldwatch Papers

On Climate Change, Energy, and Materials

On Ecological and Human Health

On Economics, Institutions, and Security

On Food, Water, Population, and Urbanization

To see our complete list of Papers, visit www.worldwatch.org/taxonomy/term/40

Price of each Paper is $9.95 plus S&H.

About the Worldwatch Institute

RESEARCH PROGRAMS

The Worldwatch Institute's interdisciplinary approach allows its team of researchers to explore emerging global issues from many perspectives, drawing on insights from ecology, economics, public health, sociology, and a range of other disciplines. The Institute's four research teams focus on:

- People
- Nature
- Energy
- Economy

PRESS INQUIRIES

Worldwatch provides reporters from around the world with access to the Institute's extensive research and the researchers behind it. For current information available to the media, visit our online press center at www.worldwatch.org/press.

For press inquiries or to be placed on the Worldwatch media list, contact Julia Tier by phone at 202-452-1992, ext. 594; by fax at 202-296-7365; or by e-mail at jtier@worldwatch.org

SPEAKERS BUREAU

Worldwatch researchers have extensive experience in bringing audiences up to date on important global trends, including food, water, pollution, climate, forests, oceans, energy, technology, and environmental security.

For more information, or to schedule a speaker, contact Darcey Rakestraw by phone at 202-452-1992, ext. 517, or by e-mail at drakestraw@worldwatch.org.

INTERNATIONAL PUBLISHING PROGRAM

Worldwatch works with overseas publishers to translate, produce, and market its books, papers, and magazine. The Institute has more than 160 publishing contracts in over 20 languages. A complete listing can be found at www.worldwatch.org/foreign/index.html.

For more information, contact Patricia Shyne by phone at 202-452-1992, ext. 520, by fax at 202-296-7365, or by e-mail at pshyne@worldwatch.org.

WORLDWATCH ONLINE

The Worldwatch Web site (www.worldwatch.org) provides immediate access to the Institute's publications. Save time and money by ordering and downloading Worldwatch publications in pdf format from our online bookstore. The site also includes press releases, special briefings on breaking environmental news, contact information, and job announcements.

SUBSCRIBE TO WORLDWATCH NEWS

Worldwatch maintains a free one-way e-mail list to distribute updates from the Institute as well as press releases on new books, papers, and magazine articles.

To subscribe, visit the Worldwatch Web site at www.worldwatch.org.

FRIENDS OF WORLDWATCH

The Worldwatch Institute is a 501 (c)(3) non-profit organization. We rely on gifts from individuals and foundations to underwrite our efforts to provide the information and analysis needed to foster an environmentally sustainable society.

Your gift will be used to help Worldwatch broaden its outreach programs to decisionmakers, build relationships with overseas environmental groups, and disseminate its vital information to as many people as possible through the Institute's Web site and publications.

To join our family of supporters, please call us at 202-452-1992, ext. 530. You can also donate online at www.worldwatch.org/donate.

LEGACY FOR SUSTAINABILITY

You can make a lasting contribution to a better future by remembering Worldwatch in your will. If you are interested in naming the Institute in your will, please contact us.

For further information on giving to Worldwatch, please contact Georgia Sullivan by phone at 202-452-1992, ext. 522; by fax at 202-296-7365; or by e-mail at gsullivan@worldwatch.org.